D1062170

Demand Forecasting and Inventory Control

Demand Forecasting and Inventory Control

A computer aided learning approach

Colin D. Lewis

JOHN WILEY & SONS, INC.
New York • Chichester • Weinheim • Brisbane • Singapore • Toronto

This book is printed on acid-free paper. ♾

Published by John Wiley & Sons, Inc.

Published simultaneously in Canada.

ISBN 0-471-25338-3

Printed in the United States of America.

10 9 8 7 6 5 4 3 2 1

To Maven and Emma and in memory of Mark

Contents

Preface xi

Part I Forecasting **1**

1 Demand forecasting and inventory control 3
Introduction 3
Forecasting versus prediction 5
Different types of forecasting methods 6
Forecast horizon, fitting and forecasting 7
Characteristics of customer demand patterns requiring forecasting 8
Inventory control 11
Conclusion 17
The OPSCON CAL module 18
Packages within OPSCON associated with this chapter 18

2 Short-term forecasting for stationary demand situations 21
Introduction 21
Forecasting in stationary demand situations 21
Statistics for establishing the validity and accuracy of forecasting models 27
The response of the simple exponentially weighted average forecasting model to different demand characteristics 29
Monitoring forecasting systems 33
Adaptive forecasting 36
Conclusion 40
Files from OPSCON's package FOREMAN associated with this chapter 41

3 Short-term forecasting for growth and seasonality 44
Introduction 44
Growth forecasting models 44
Seasonal forecasting 48
Conclusion 51
File from OPSON's package FOREMAN associated with this chapter 52

4 Medium-term forecasting for growth and seasonality 54
Introduction 54
Regression and time series analysis 55
Regression statistics for establishing the validity and accuracy of forecasting models 59
Time series analysis – forecasting: seasonality 60
Conclusion 64
Files from OPSCON's package FOREMAN associated with this chapter 64

5 The essential links between forecasting and inventory control 68
Introduction 68
The essential links between forecasting and inventory control 68
Selection of forecasting models based on type of stocked item – Pareto analysis 69
Choosing the most appropriate group of forecasting model(s) for stocked items 72
Conclusion 76
File from OPSCON's package STOCKMAN associated with this chapter 76

Part II Inventory control – the re-order level inventory policy 79

6 Establishing the value of the re-order level 81
Introduction 81
The re-order level (ROL) policy 81
The traditional approach to establishing the value of the re-order level 84
Linking the forecast of demand to the setting of the re-order level 89

Conclusion 90
File from OPSCON's STOCKMAN associated with this chapter 90

7 Establishing the size of the replenishment order 94
Introduction 94
The EOQ approach to establishing the value of the replenishment quantity 94
Criticisms of the EOQ approach 98
Conclusion 104
Files from OPSCON's package STOCKMAN associated with this chapter 105

8 Examining the relationship between the re-order level and the replenishment quantity 108
Introduction 108
Interpreting the service level based on simple statistical concepts 108
Customer service level defined as the proportion of annual demand met ex-stock per annum 111
Conclusion 115
File from OPSCON's package STOCKMAN associated with this chapter 116

Part III Inventory control – the re-order cycle inventory policy 117

9 Establishing the review period and maximum stock level 119
Introduction 119
Selecting the review period in the re-order cycle inventory policy 119
Setting the maximum stock level 121
Establishing the customer service level defined as the proportion of annual demand met ex-stock per annum within a re-order cycle policy 124
Linking the forecast of demand to the setting of the maximum stock level 125
Conclusion 126
File from OPSCON's package STOCKMAN associated with this chapter 126

Part IV Conclusion **131**

10 Selecting the most appropriate inventory control policy 133
Introduction 133
Selection of inventory control policy based on type of stocked item – Pareto analysis 134
Hybrid or specialised inventory policies 136
Conclusion 140
File from OPSCON's package STOCKMAN associated with this chapter 141

Appendix A The OPSCON package 144

Appendix B Simplification of exponentially weighted average formula 152

Appendix C Selected values of the normal variable, u 153

References 154

Index 155

Preface

This book allows both students and practitioners to understand the theory and practice of current demand forecasting methods, the links between forecasts produced as a result of analysing demand data and the various methods by which this information, together with stocked item cost information, is used to establish the controlling parameters of the most commonly used inventory control systems.

The demand forecasting section of the book concentrates on the family of short-term forecasting models based on the exponentially weighted average and its many variants and also a group of medium-term forecasting models based on a time series, curve fitting approach. Although the forecasting models described are chosen primarily for their proven ability to forecast future demand, they can be used to analyse any variable whose values occur at regular intervals of time.

The inventory control sections of the book investigate the re-order level policy (or fixed order quantity system) and re-order cycle policy (or periodic review system) and indicate how these two policies, provided with the appropriate forecasting and cost information, can be operated at minimum cost while offering a high level of customer service. Consideration is also given to the hybrid (s,S) inventory control policy and to the situation characterised by items such as slow moving spares, where the demand for a single unit can be expressed by an average interval between issues as long as twenty years.

Which of the many demand forecasting models and inventory control policies available is most appropriate for particular stocked items is discussed in the context of a Pareto analysis approach whereby such items are allocated into 'A', 'B' and 'C' categories based on annual usage value.

The accompanying integrated, computer aided learning (CAL) package OPSCON (OPerationS planning and CONtrol) allows readers to create demand data files or use prepared data files for subsequent forecasting analysis. Following this forecasting analysis, a unique feature of the package allows the user, on terminating the forecasting analysis within the forecasting package FOREMAN, to 'carry' the resulting forecast information as demand input to the inventory control package STOCKMAN's two major inventory control policies. These two policies are represented by Monte Carlo simulations of what would have occurred in practice over a twelve year period assuming an imposed monthly demand and cost environment. As a result of the simulation analysis, recommendations are made for those values of the parameters controlling the chosen inventory policy which would offer a high level of service to customers at a minimum cost. Since all the necessary calculations are performed automatically, for the practitioner in particular, the OPSCON package when provided with demand and cost information can be used to establish the correct inventory control parameters without the user necessarily having acquired any knowledge of the underlying theory.

The integrated CAL package OPSCON occupies about 1.2 Mbytes of disk space and can be installed on and run either from the hard disk of any IBM-compatible PC within a DOS or WINDOWS environment or on a local area network (LAN). The package is totally menu driven with each menu composed of options with unique first capitalised (upper case) characters making each choice available with a single key depression. This allows the package to be used just as easily by the uninitiated as those who have studied the accompanying text. The installation and operating instructions for OPSCON for those who can't wait to get started are in Appendix A!

COLIN LEWIS

Part I

Forecasting

1

Demand forecasting and inventory control

Introduction

Forecasting is a necessary pre-requisite to most operational activities. Without an estimate of the future it is not possible to plan for the level of activity which is to be expected and, hence, not possible to estimate the resources that need to be designed, planned and controlled to fulfil that level of activity.

Inventory control is the science-based art of controlling the amount of inventory (or stock) held, in various forms, within an organisation to meet the demand placed upon that business economically. To control the amount of inventory, it is clearly necessary to forecast the level of future demand, where such demand can be regarded as essentially either independent or dependent.

Independent demand describes the type of demand for a product or service which is independent of demand for other apparently related products or services. Many services organisations, public utilities and retail organisations display such independent demand characteristics. Within a supermarket, for instance, it is unlikely that the demand for bread is dependent on the demand for ice cream. Also, even within manufacturing environments – where demand at lower levels of the production process are clearly dependent on demand at higher levels – from a stock holding point-of-view it may be more sensible to ignore this self-evident dependency and assume that trends identified in past data – using a forecasting procedure – could be a more reliable (and certainly a more cost effective) indicator of future demand (i.e. that it is assumed that demand is independent). This apparent conundrum will be explained later in this chapter.

However, generally within manufacturing organisations, where a planned level of production of finished products is the norm, the demand for sub-assemblies, components and raw materials that make up the finished product is clearly highly dependent on that planned level of production. Hence in automobile manufacture, the demand for wheels, tyres, etc, is essentially a multiple of five times the demand for completed vehicles (i.e. four road wheels plus a spare). The methods of controlling the provisioning of sub-assemblies, components and raw materials for such dependent demand situations are radically different from those for independent demand.

Within a typical dependent demand environment, given a commitment to a future production plan of a specified number of finished products at specified times (Master Production Schedule) and also given that information is available such that the number of sub-assemblies, components and raw materials that make up each finished product can be defined (Bill of Material), combined with information as to the current stock holding situation of these items (Status of all Materials), the future demand for sub-assemblies, components and raw materials in terms of their required quantities and timing can theoretically be forecast with a high degree of certainty and accuracy. The process which converts a production plan of finished products into a schedule of supply of sub-assemblies, components and raw materials is referred to by the generic term Material Requirements Planning (MRP) which is essentially a large, and often very complicated, database manipulation exercise which requires large amounts of information regarding the relationships between finished products and their constituent sub-assemblies, components and raw materials. Given this information, which also needs continually updating if any engineering changes are introduced, demand at all levels of the production process can theoretically be established by exploding the planned demand of finished products at the highest level into sub-assemblies, components and raw materials at progressively lower levels and aggregating the results over all levels of the production process, while also taking into account existing stocks. Although simple in concept, the successful application of MRP to a production schedule of a mix of finished products within an existing inventory situation and subject to the vagaries of the production processes involved and the in-built inaccuracies of the information held in the Bill of Material is a target for which many manufacturing companies aim, but few achieve completely.

Given that such a dependent demand process could establish future demand at all levels of production with complete accuracy, in theory there would no need for any stocks of sub-assemblies, components or raw materials since their delivery could be synchronised to coincide exactly with their requirement. Hence the adoption of terms such as Just in Time (JIT) and Zero Inventory. The reality however is that in practice such an ideal situation is virtually impossible to achieve, although the best practitioners claim to come close.

Within a dependent demand environment, at the lower levels in such a production process, rather than relying on an expensive and complicated database exercise which involves the process of exploding and aggregating demand over several levels of production, if past demand can be assumed to be a reasonable indicator of future demand (i.e. demand can be assumed to be independent) stocks held by traditional inventory control policies with parameters established through a forecasting regime may well be cost effective. Certainly no automotive manufacturer establishes the demand for the many standard nuts, bolts, washers and fasteners through an MRP process – even though technically it should be possible. Even for such relatively cheap items, however, it is still necessary for the inventory control policies which are used to control such items do so effectively with a high level of service but at a reasonable cost.

This book and its accompanying CAL package OPSCON are concerned solely with the control of items whose demand characteristics are independent. Readers whose area of interest covers dependent demand situations are recommended to consult Harrison's[2] *Finite capacity scheduling – the art of synchronized manufacturing* and Kenworthy's[5] *Planning and control of manufacturing processes.*

Forecasting versus prediction

In general terms, forecasting is interpreted as being a scientific process of estimating a future event by casting forward past data. The past data are initially analysed to establish the underlying trends which characterise the data and this information is then used in a predetermined way to obtain an estimate of the future. Prediction, however, is generally interpreted as a process of estimating a future event based on subjective considerations. However, a scientifically produced forecast, established on the assumption that characteristic trends identified in past data will continue into the future, should always be open to alteration in the light of changes in market conditions. Examples of such changes could be:

- it is predicted, in advance of the event, that such trends are unlikely to be continued due to legislative changes which are assumed to affect future demand, or
- after the event, some extraneous causal effect is detected which invalidates the assumption of continuity of demand.

However, because predictions are predominantly subjective, they are generally more expensive to produce on a routine basis than forecasts. Hence, where many product lines are involved, it is generally more effective to operate on the assumption that scientifically produced forecasts are assumed to be satisfactory unless a monitoring procedure indicates that the forecast for a particular product line or item is no longer in control, at this time predictions of future demand may well be required to re-establish the validity of the forecast.

Different types of forecasting methods

One way of classifying or categorising forecasting methods is to define the type of forecast on the basis of the time period associated with the demand data which are being analysed, as illustrated in Table 1.1.

Short-term forecasting

Although there is no strict demarcation between the various forms of forecasting categorised within Table 1.1, it is generally assumed that short-term forecasting will often be associated with many product lines or items, as typically occurs in an inventory control environment. Within an inventory environment, it is also true that the demand patterns being analysed are relatively fast moving with an average per period in excess of twenty such that the normal probability distribution can be assumed to represent the distribution of demand per unit time. The forecasting models used when operating in such an environment must be simple and relatively cheap to operate while still being robust. The family of forecasting models based on the exponentially weighted average, originally suggested by Holt,[3] has proved to meet these criteria more satisfactorily

Table 1.1 Categorisation of type of forecast based on the underlying time unit of data involved

Category of type of forecast	Time period associated with the data being analysed	Example of forecasting application	Forecasting techniques used
Immediate-term	¼ hr to 1 day	Electricity demand forecasting	Various
Short-term	1 week to 1 month	Demand forecasting in industry and commerce	Exponentially weighted averages and derivatives
Medium-term	1 month to 1 year	Sales and financial forecasting	Regression, curve fitting, time series analysis
Long-term	1 year to 1 decade	Technological forecasting	DELPHI, think tanks, etc

than any other group of models and has traditionally become the basis for forecasting in many inventory control situations because of its:

- computational cheapness in terms of processing time and storage requirements;
- robustness and ability for the forecasting process to be monitored so that 'out of control' situations can be detected;
- ease of 'starting up' when including new items with no previous demand history;
- adaptability in terms of changing sensitivity in line with the characteristic of the demand data encountered; and
- flexibility in terms of ability to cope with stationary, growth and seasonal demand patterns.

Medium-term forecasting

Increasingly as computer power becomes cheaper more sophisticated forecasting models, which require more complicated calculations than the exponentially weighted average family of forecasting models, are becoming financially viable as a method of forecasting for many stocked items. Forecasting methods such as:

- curve fitting and regression;
- Fourier analysis; and
- Baysian forecasting;

now feature in some of the more advanced, professionally developed software for inventory control.

Forecast horizon, fitting and forecasting

The forecast horizon is the number of time periods ahead of the known data over which forecasts are calculated (i.e. extrapolation). In many situations the maximum forecast horizon is about six periods ahead, since the confidence with which forecasts any further ahead of this can be made is likely to be low. An exception to this general rule is when a strong seasonal influence is known to exist, in which case forecasts up to a whole season ahead might well be justified i.e. up to twelve months for monthly data.

Fitting is the process of producing a forecasting model which fits known data (i.e. interpolation) whereas forecasting is the process of extrapolating a fitted model into the future, i.e. ahead of known data.

Within an environment where forecasting models are based on parameters which can be adjusted, a good fit is usually established by adjusting the value of those parameters to minimise the Sum of Squared forecasting Errors (SSE) or the Mean Squared forecasting Error (MSE).

It is generally assumed that the best fitted model will also be the best model for forecasting where, in its strict interpretation, forecasting produces future estimates ahead of known data.

Characteristics of customer demand patterns requiring forecasting

Inventory control systems must cope with a variety of different customer demand patterns (for which forecasts are necessary) if an effective overall policy for controlling inventory (or stocks) is to be achieved. In practice, in ascending order of complexity, it is assumed that the following demand patterns can exist:

1. **Stationary demand** – assumes that although customer demand per unit time fluctuates, there is no apparent long-term underlying growth or seasonal trend. Figure 1.1 illustrates the basic stationary character of such data but also identifies the fact that variability in demand exists. Because no growth or seasonality is assumed in stationary demand patterns, forecasts ahead are fixed in value and the forecast for one period ahead is the forecast for any number of periods ahead.

   Although within a stationary demand pattern no growth or seasonality is assumed to exist, it should be accepted that occasionally fundamental changes in the demand pattern may occur, but these are presumed to be short-term in nature, such as:

   - impulses – individual demands which are significantly higher or lower than normal. Such impulses are best ignored by a forecasting system linked to an inventory control policy, since such policies are basically designed to cope with a reasonably stable level of demand with a known, measurable degree of variation; and
   - step changes – a series of successive demands which are significantly higher or lower than normal. The ideal response of a forecast to a step change in demand is that it should react as quickly as possible. Should this not be feasible, a competent forecasting system should at least identify that such a step change has occurred and should also instigate remedial action to ensure that the forecast, which will naturally lag behind such a sudden change of level, is corrected.

   Figure 1.2 illustrates a demand pattern where a single period

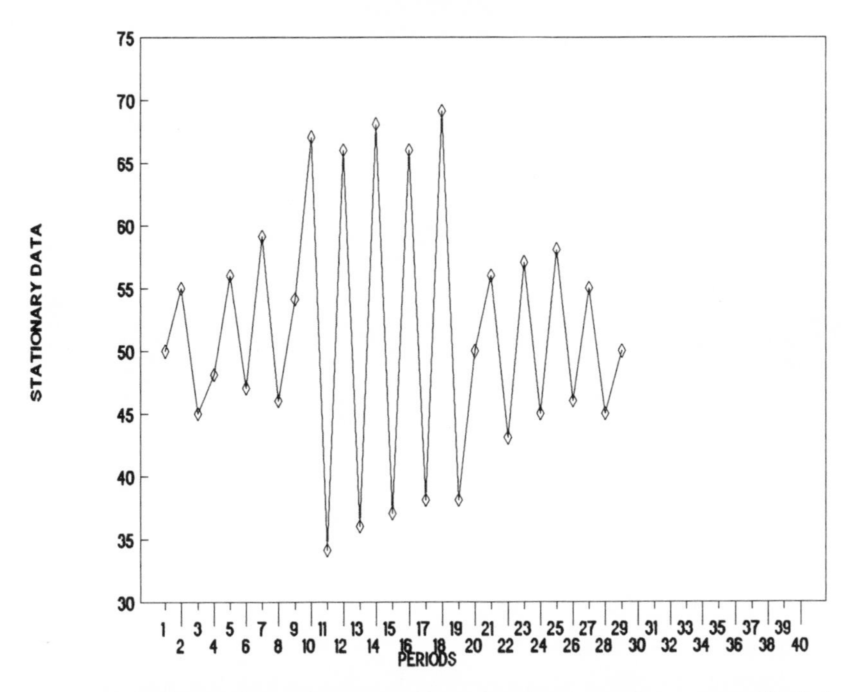

1.1 Stationary demand pattern with no long-term growth or seasonality but with considerable variability of demand. This data series is available within OPSCON as STATIONARY.

impulse (a significant, high demand occurring for one period only) is followed by a positive step in demand (a succession of significantly high values).

2. **Demand with growth characteristics** – where a demand pattern exhibits a growth characteristic over the longer term, the forecasting models required to accommodate that growth become more complex than those used in the stationary demand situation discussed earlier. Such inappropriate stationary forecasting models produce fitted forecasts which lag behind the data together with forecasts ahead which are fixed in value. Figure 1.3 illustrates a demand pattern where demand is growing long term.

   NOTE: It is recognised that demand may decline rather than grow, in which case this continual drop in demand can be regarded as negative growth.

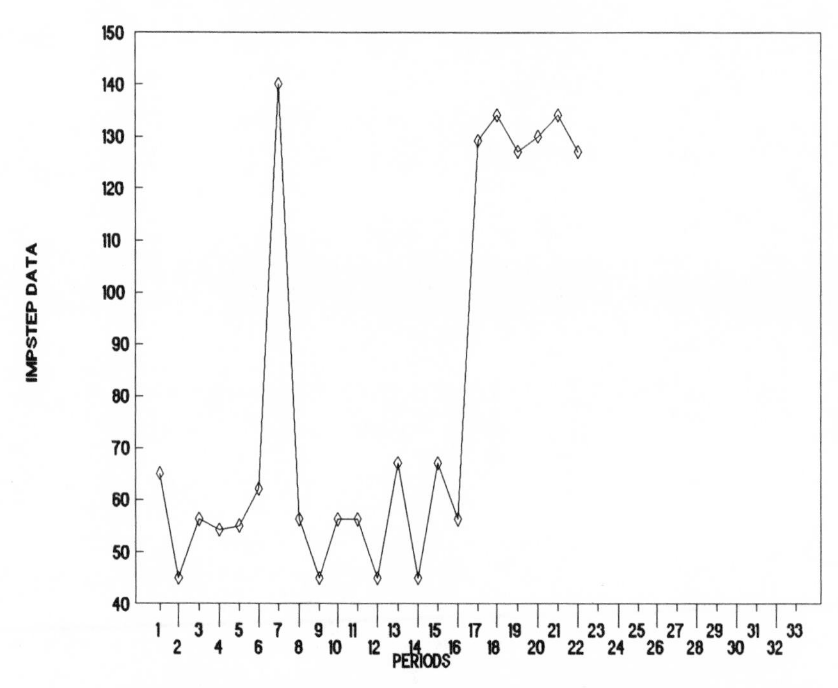

1.2 A stationary demand pattern where an impulse is followed by a change step. This data series is available within OPSCON as IMPSTEP.

3. **Demand with seasonal characteristics** – many demand series are influenced by the seasons of the year and by other events which occur annually. In such situations it is possible to establish the degree to which demand in any particular period of the year (i.e. month, quarter or accounting period) is higher or lower than for a typical average period. Hence the aim of forecasting models taking seasonality into account is to establish this relationship for each and every period within the year and to use the de-seasonalising factors that are identified by this process to produce forecasts. For technical reasons it is generally assumed that growth may also exist in demand patterns characterised by seasonality as is shown in Fig. 1.4. If there is no growth, the analysis simply registers actual growth as negligible.

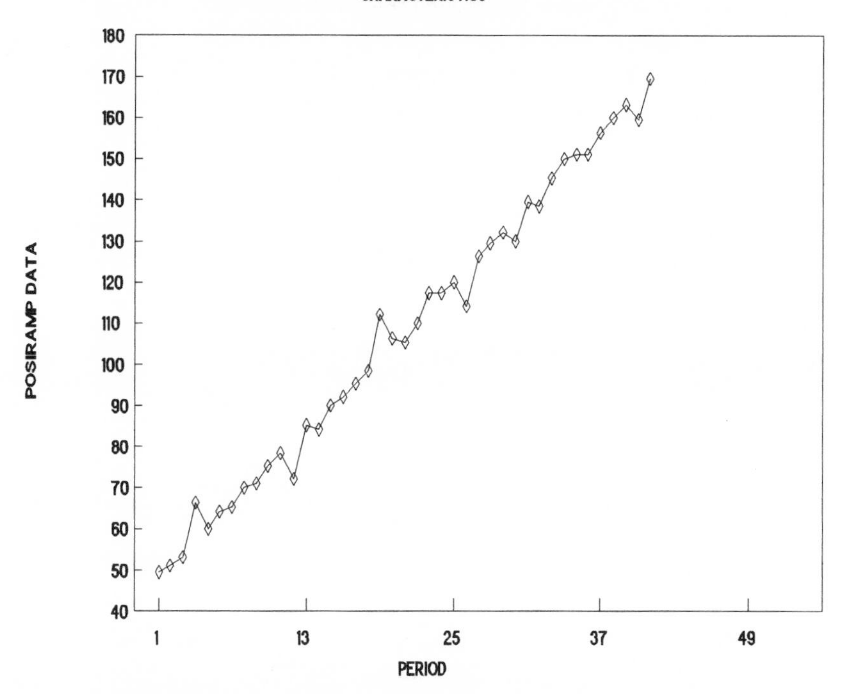

1.3 A demand pattern characterised by long-term growth. This data series is available within OPSCON as POSIRAMP.

Inventory control

The benefits of holding stock

To be viable, the holding of inventory (or stocks) must offer benefits to the customers whose demand requirements are met by the stock holding process. Some of the principal benefits of holding stock are:

1. The demands of customers are buffered by the stock holding process from the suppliers of replenishment orders to the stock holding process. Thus, given that the holding of stock is organised economically, the frequent small demand orders of customers can be met more effectively by the supply of large, infrequent replenishment orders by suppliers. Thus stock can often act as an effective cushion against interruptions to supply. Although not immediately thought of in terms of an item

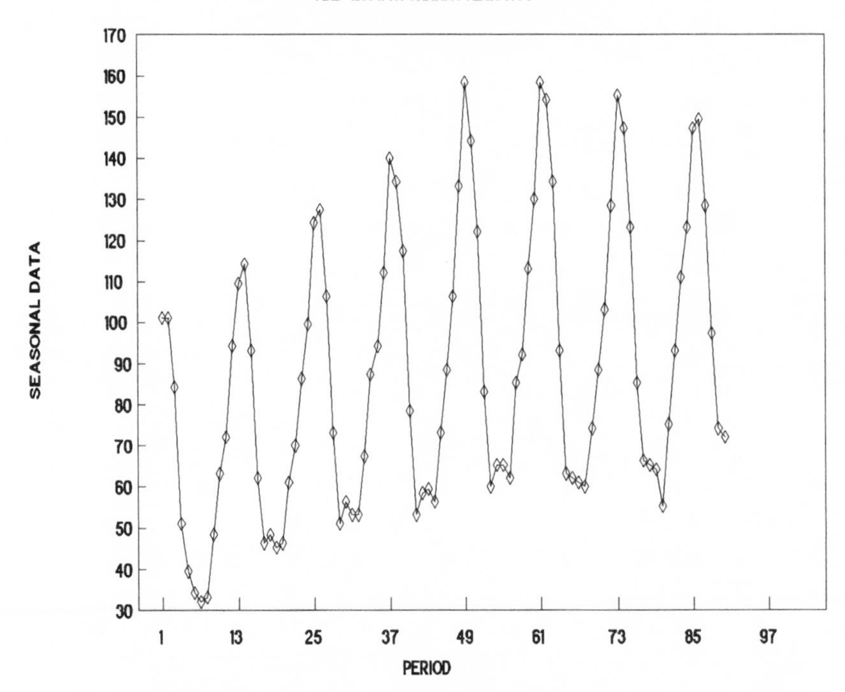

1.4 A demand pattern characterised by long-term growth and seasonality. This data series is available within OPSCON as SEASONAL.

controlled by orthodox inventory control procedures, stocks of drinking water facilitated by the construction of reservoirs clearly assist in the process of meeting customers' demand throughout the year, even though the supply of water through natural rainfall is effectively temporarily interrupted during the summer months.

2. Because inventory systems are replenished with relatively large orders, the unit costs of the items provisioned by such large orders can often be reduced due to:

- longer production runs absorbing set-up costs more effectively;
- discounts or price breaks offered for large orders; and
- more economical transport and packaging costs.

Inventory operating costs

To offset the benefits of holding stocks, it must always be recognised that the stock holding process incurs specific operating costs, the more important of which are:

1. Ordering costs – which include the costs involved in raising replenishment orders which may involve the paperwork involved together with the associated labour costs. These can be reduced drastically through computerisation and the use of EDI (Electronic Data Interchange) but are never negligible.
2. Storage (or holding) costs – which are generally expressed as a percentage of the unit cost of the item involved (typically 20–30% pa) and made up of:

    - the costs of borrowing money to invest in stock or alternatively the loss of possible interest that could have been made had a company not used its own money to invest in stock (typically 10–15% pa);
    - the costs directly associated with storing goods, i.e. storemen's wages, rates, heating and lighting, store's transport, racking and palletisation, protective clothing, weighing equipment, etc;
    - costs involved in preventing deterioration of stock; (typically 0–30% pa);
    - costs incurred as a result of obsolescence which could include costs of re-work or scrapping (typically 4–7% pa); and
    - costs of insurance to cover fire, theft and third party injury (typically about 1.5% pa).

3. Stockout costs – costs which are incurred as a result of not being able to satisfy customer demand when stocks are not available. These costs are generally very difficult to define in terms of whether they should be assessed on a basis of the actual number of units of shortage or the time period over which the shortage occurs. In view of this difficulty, most inventory control models ignore stockout costs and argue that they are provided for, if somewhat obliquely, by offering a reasonable level of service to customers of the inventory system.
4. Administration costs – which can cover a multitude of fixed costs which are only incurred because stocks are held.

Inventory control policies

To run a successful inventory system, which should effectively balance the supply of replenishment orders against the requirements of demand, it is clearly necessary to have a policy which defines a series of rules as to how and when replenishment orders are to be raised. Essentially the raising of replenishment orders can be triggered either when stock-on-hand (the physical stock held plus any outstanding replenishment orders less committed stock) reaches a certain level or alternatively replenishment orders can be raised on a regular time basis i.e. once a month, once a quarter, etc.

If replenishment orders are raised on a level basis, to maintain reasonably balanced stocks, the size of the replenishment order is usually fixed in size leading to a policy known as the re-order level (or re-order point) policy within which both the value of the re-order level has to be specified (see Chapter 7) together with a specification of the size of the replenishment quantity (see Chapter 8).

Alternatively, if replenishment orders are placed on a regular time basis, because the level of stocks at the time the replenishment order is placed will vary considerably, it is usual to place variable size replenishment orders; larger orders being placed when stock levels at review are low and conversely smaller replenishment orders when stock levels at review are high. The simplest and most effective method of arranging such a replenishment ordering policy is to raise an order whose size is established on the basis of a specified maximum stock level less the level of stock held at the time of review. Such a policy is referred to as a re-order cycle policy (see Chapter 9).

Both the re-order level and re-order cycle policies have advantages and disadvantages and these have to be considered in detail when deciding which type of policy is most appropriate for a particular stocked item. However, whichever policy is chosen, it is essential that the parameters controlling the operation of the policy (re-order level and replenishment quantity in the case of the re-order level policy and maximum stock level and review period in the case of the re-order cycle policy) are set at the correct levels such that the policy offers a reasonable level of service in terms of meeting demand without doing so at excessive cost.

Such control parameters are clearly going to be dependent on the characteristics of the demand imposed on the stock control system and the delay in replenishment orders being delivered (leadtime). This is demonstrated clearly by graphs of stock balances generated by the OPSCON package as shown in Fig. 1.5 to 1.7.

In Fig. 1.5 it can be seen that, when controlling stock with a re-order level policy, the stock balances displayed indicate that the policy appears to be operating reasonably effectively in that enough stock is held so that stockouts (periods when demand cannot be satisfied) are likely to be relatively few while on the other hand stocks never become excessively large. In this situation the average demand per month has been established

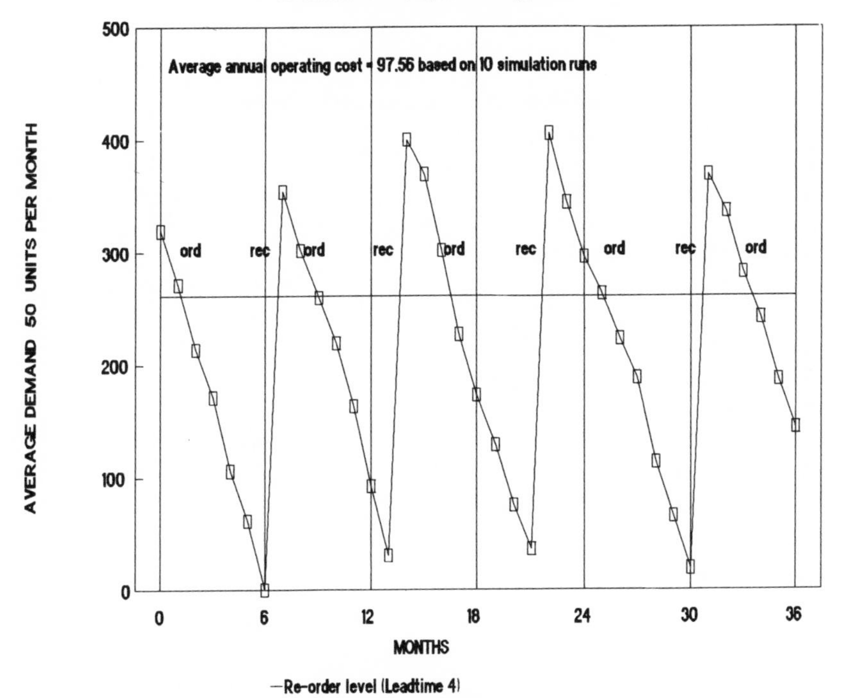

1.5 A correctly tuned stock control policy showing stock balances satisfying demand at minimum cost.

at 50 units. For a particular stocked item, and operating the policy with a re-order level of 260 and a replenishment quantity of 400 subject to a delivery leadtime of 4 months this produces an annual average total operating cost of 97.56. Clearly the parameters controlling this inventory policy have been well chosen in the light of the demand and replenishment leadtimes imposed on the system, and the total annual operating costs (storage, ordering and stockout costs) can be assumed to be as low as is possible in this demand environment.

Figure 1.6 shows the stock balances for the same re-order level inventory policy operating under the same controlling parameter values as shown in Fig. 1.5, but now in a situation where the average demand per unit time has increased by 20%. Here, the increase in demand has led to extremely low stocks, which has inevitably created frequent stockouts and hence a significant increase in stockout (or penalty) costs. These increased penalty costs have not been compensated for by a relatively small drop in

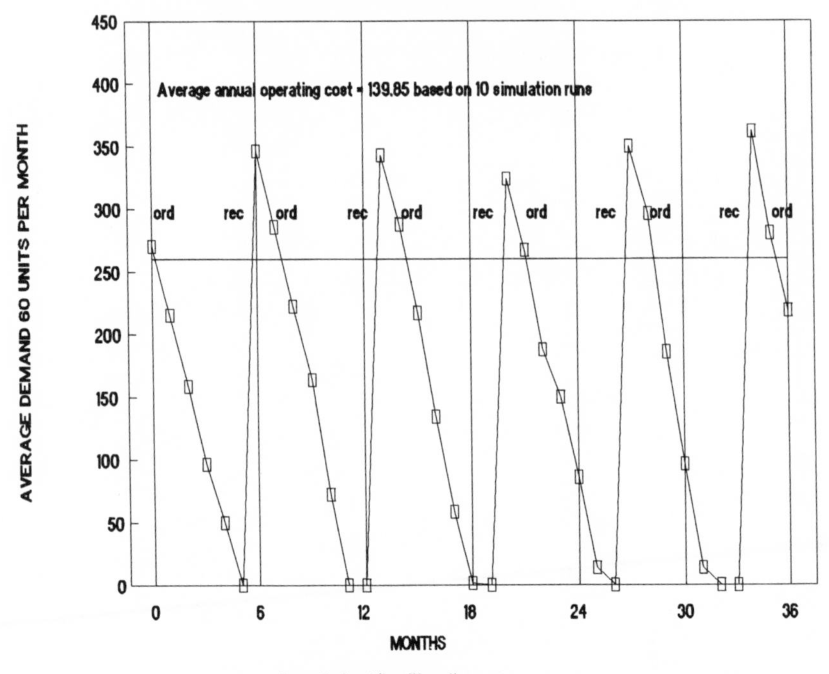

1.6 An incorrectly tuned stock control policy where a 20% increase in demand has not been accommodated by a resetting of the controlling parameters. This has led to a shortage of stock and resulting stockouts.

storage costs, resulting in an overall increase in the annual average total operating cost from 97.56 to 139.85.

To regain a minimum cost operation in this new situation, it would be necessary to alter the values of the parameters controlling the inventory policy (say by increasing the re-order level and the replenishment quantity) in the light of the increased demand. Figure 1.7 demonstrates this situation, where to compensate for the increased demand the re-order level has been raised from 260 to 300 units and the replenishment quantity from 400 to 438 units. The resulting average annual operating cost in this situation is 107.72, significantly lower that the 139.85 incurred in the previous situation.

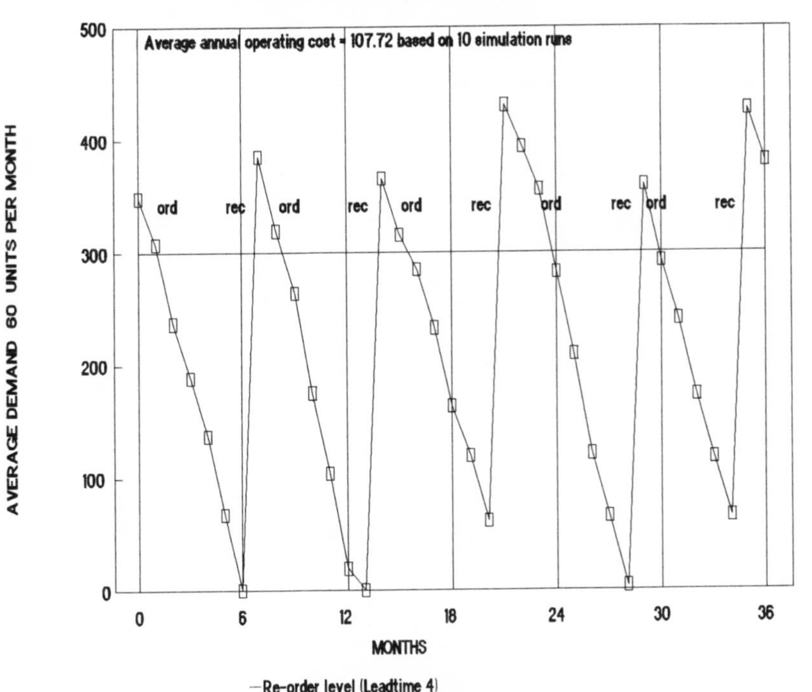

1.7 Stock balances for a stock control policy where the controlling parameters have been correctly adjusted to compensate for a 20% increase in average demand.

Conclusion

In more general terms, it should now be apparent that the parameters controlling inventory polices are:

- for the re-order level policy, the re-order level value and the size of the replenishment order; and
- for the re-order cycle policy, the review period and the maximum stock level used to evaluate replenishment quantity values.

The setting of these parameters must be linked to the level of demand imposed on the inventory system and also to the delay between placing replenishment orders and their subsequent receipt (i.e. the leadtime).

This relationship between forecasting the level of demand and using such forecasts together with estimates of the replenishment leadtime to adjust (or fine tune) the controlling parameters of inventory policies is the essence of inventory control theory and thus of this book and its associated CAL package.

The OPSCON CAL module

The OPSCON CAL module contains the two inter-related packages FOREMAN and STOCKMAN. The OPSCON module can be run on either a standard PC by a HARD-DISK USER or on a network by a NETWORK USER.

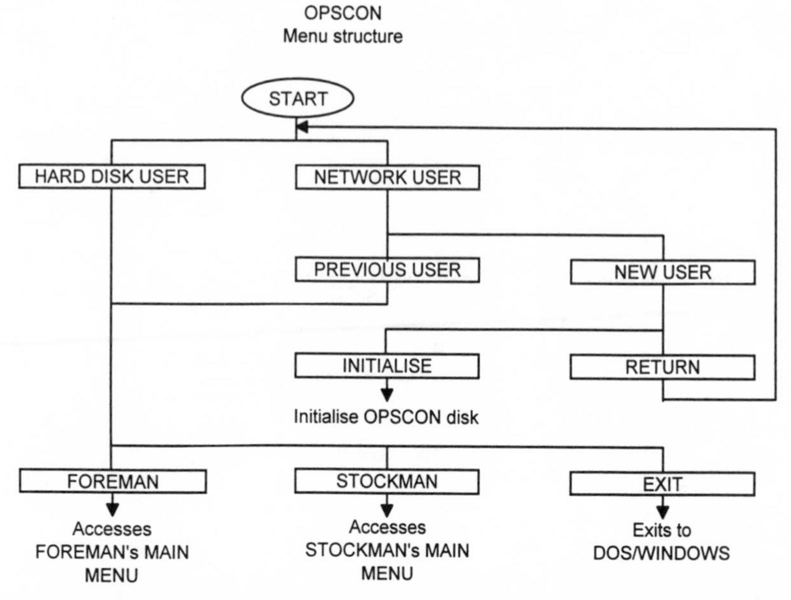

For network users it is assumed that all temporary files to be saved by the OPSCON module, and also graphical .PIC files specified by the user, will be saved to a floppy disk located in the A:\ drive. The disk used for this purpose needs to be initialised and is subsequently referred to as the OPSCON disk.

Packages within OPSCON associated with this chapter

FOREMAN package – contains six files for analysing demand series and producing forecasts as follows:

- EWA – a file detailing the calculations involved in the simple exponentially weighted average, adaptive and delayed adaptive response rate forecasting models;

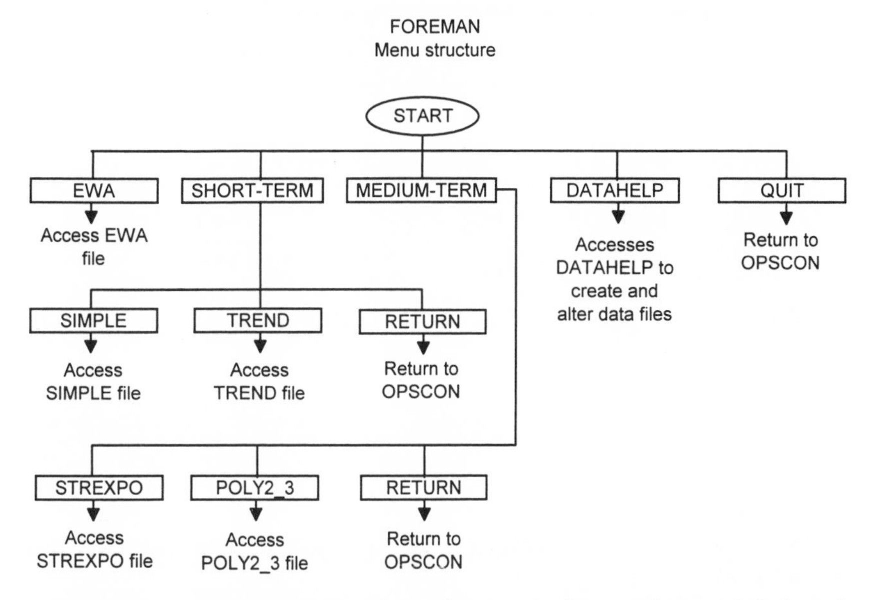

- SIMPLE – a file containing three short-term forecasting models based on the simple exponentially weighted average, adaptive and delayed adaptive response rate methods. Any demand series can be analysed and three sets of DEMONSTRATION data are provided;
- TREND – a file containing three short-term forecasting models based on Brown's double smoothed exponentially weighted average and Holt Winters' quarterly and monthly seasonal methods. Any demand series can be analysed and three sets of DEMONSTRATION data are provided;
- STREXPO – a file containing two medium-term forecasting models based on a straight line and exponential growth trend curves. Quarterly and monthly seasonal analysis can be initiated once a trend has been chosen. Any demand series can be analysed and three sets of DEMONSTRATION data are provided;
- POLY2_3 – a file containing two medium-term forecasting models based on second and third order polynomial trend curves. Quarterly and monthly seasonal analysis can be initiated once a trend has been chosen. Any demand series can be analysed and three sets of DEMONSTRATION data are provided; and
- DATAHELP – a file providing facilities for creating and altering data files suitable for analysis by the forecasting files SIMPLE, TREND, STREXPO and POLY2_3.

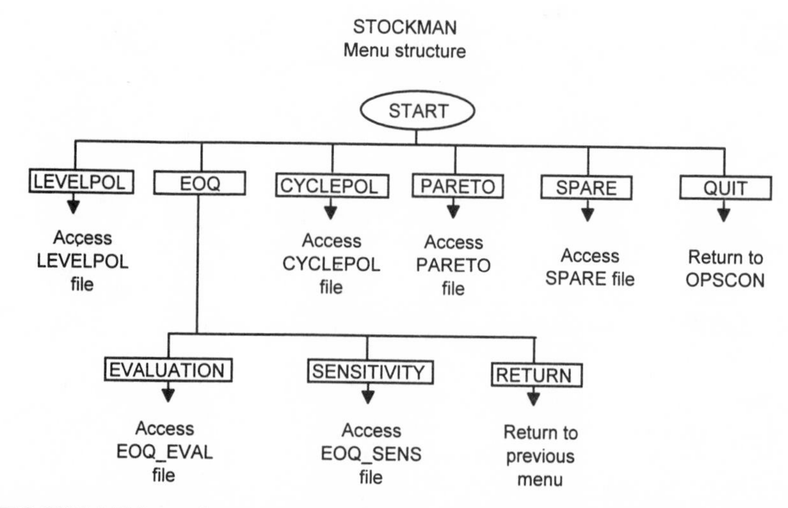

STOCKMAN package – contains six files for simulating inventory control policies and demonstrating relationships within inventory control, as follows:

- LEVELPOL – a file which simulates the operation of a re-order level inventory policy. Options allow the user to change policy parameters, view resulting graphs and alter costs;
- EOQ_EVAL – a file which progressively develops the formulation of the Economic Order Quantity. Once the formulation is complete cost and demand variables can be altered to study the effect on the EOQ;
- EOQ_SENS – a file which compares numerically and graphically the costs involved in operating at a replenishment order quantity other than the EOQ;
- CYCLEPOL – a file which simulates the operation of a re-order cycle inventory policy. Options allow the user to change policy parameters, view resulting graphs and alter costs;
- PARETO – a file consisting of 250 stocked items which can be sorted either alphabetically by part number or in descending order of value to demonstrate the Pareto or ABC relationship; and
- SPARE – a file demonstrating the costs involved for an inventory control situation of expensive, slow moving spares where the demand may be as low as only once every twenty years.

2

Short-term forecasting for stationary demand situations

Introduction

The choice of which forecasting method is most appropriate when linked to an inventory control system is predominated by the likely number of different forecasts that have to be made on a routine basis. For many inventory systems, this could be several thousands. Hence, a high degree of sophistication is clearly not required and a family of forecasting models which are cheap to operate but which provide reasonably accurate forecasts is to be preferred. The exponentially weighted average family of forecasting models is one group that fits this criterion most satisfactorily and since its first exposure by Holt[3] has been a popular forecasting model selection when linked to inventory control. This chapter details the simple exponentially weighted average forecasting model and two of its derivatives and contrasts their features with the more commonly known moving average and then considers for which type of demand data this model is most suited.

Forecasting in stationary demand situations

The simplest demand environment within which to produce forecasts occurs when it can be assumed that the underlying demand process is stationary (see Chapter 1, page 8) within which no growth or seasonality is assumed. The basic inference within a stationary demand process is that there is variation about a relatively stationary average demand value and that any change in the average value (i.e. a movement upwards or downwards) is due to a special, one off cause rather than to overall growth or seasonality. In this chapter this simplest of demand processes is discussed and later chapters then cover the more complex demand patterns where both growth and seasonality are considered to exist.

Relating forecast and demand data to time periods

Before developing specific forecasting models to be linked with inventory control policies, it is clear that in all forecasting situations it is necessary to define the timing of forecasts and demand data to the particular time period to which they belong or relate. The convention adopted is to regard the current period (i.e. now) as present time, t, and refer all other timings to present time. In practice, therefore, d_t defines the demand that occurred in the most recent period under consideration. Hence for the changeover from one period to the next – at the end of the previous period and the start of a current period – d_t could be interpreted as being 'previous period's demand' since this is the most recent demand value which has occurred.

Past time is considered as negative with respect to the current period, t, hence d_{t-1} defines the demand that occurred in the period immediately previous to the period in which d_t occurred. Using the same definition for the current period as before, d_{t-1} could be interpreted as being 'the period-before-the-previous period's demand'.

Although demand data can only occur in the past, forecasts are targeted at the future. Hence, future time is defined as positive with respect to the current period and f_{t+1} would define the timing of the forecast for the period following the current period. At the start of a new period defined as occurring at time t, f_{t+1} could be interpreted as the 'forecast for the period one-period ahead of the current period'.

Assuming that the demand value has just been collected for the month of March, and that a forecast is required for April, the timings in Table 2.1 would apply. In a stationary demand situation, because no growth or seasonality is assumed, the forecast for one period ahead is the forecast for any number of T periods ahead, where T is any specified forecast horizon. Hence, in the stationary demand situation *only* the forecast for T periods ahead f_{t+T} is given by:

$$f_{t+T} = f_{t+1}$$

Note: Within the OPSCON package it is assumed that time periods can be either calendar months (twelve per year) or weeks (fifty per year)

Table 2.1 Relating demand values and forecasts to time periods

Month	Jan	Feb	Mar	Apr
Time period	(t−2)	(t−1)	(t)	(t + 1)
Demand values	40	44	36	?
Forecasts				40

The moving average as a forecasting model

Referring to Table 2.1, the forecast of 40 for next month (April) f_{t+1} based on a moving average, m_t, calculated this month (March i.e. now at time, t) could be evaluated as:

$$f_{t+1} = m_t = (1/3)d_t + (1/3)d_{t-1} + (1/3)d_{t-2} = (1/3)40 + (1/3)44 + (1/3)36 = 40$$

The more general form of the moving average as a forecasting model would be:

$$f_{t+1} = m_t = (1/n)d_t + (1/n)d_{t-1} + \dots (1/n)d_{t-n+1} \quad [2.1]$$

where n = 2,3 ... 12, etc, and where the sum of the n weights of 1/n will always sum to one; **this being the definition of a true average.**

However, in practice, the use of a moving average as a forecasting model poses the following practical problems:

- it is difficult to initialise, i.e. to start from a situation where no data exist;
- it requires relatively large amounts of data to be stored e.g. if a 12 period moving average is to be operated, 11 periods of demand values would need to be stored until the 12th period's value becomes available to complete the calculation as defined by Equation [2.1]. This represents a significant data storage problem if forecasts are to be provided for several thousand stocked items;
- it imposes difficulties in changing sensitivity in that the number of periods included within the average would need to be varied;
- it requires that all data included are weighted equally; and
- it imposes a sudden cut off in weighting for data not included, i.e. with a 12 period moving average the demand value related to the period just excluded from the average contributes absolutely nothing to that average whereas the previous period's value (only one period younger) contributes 1/12 or 8.33%

The final problem of equal weighting could be overcome by developing a one-period-ahead forecast based on an unequally weighted moving average, hence:

$$f_{t+1} = m_t = (0.5)d_t + (0.3)d_{t-1} + (0.2)d_{t-2}$$

which is a valid, unbiased forecasting model since the sum of the weights add up to one.

It is the extension of this concept of an unequally weighted moving average which led to the development by Holt[3] of the simple exponentially weighted average, an average with an infinite number of weights which decreased exponentially with time.

The simple exponentially weighted average as a forecasting model

The definition of an average, u_t, with weights declining exponentially with time would be of the general form:

$$u_t = \alpha d_t + \alpha(1-\alpha)d_{t-1} + \alpha(1-\alpha)^2 d_{t-2} + \alpha(1-\alpha)^3 d_{t-3} + \alpha(1-\alpha)^4 d_{t-4} \ldots \quad [2.2]$$

where (alpha) α is an exponential weighting constant (EWC) whose value must be between zero and one; given that the sum of weights must sum to one (at infinity). A value of $\alpha = 0.2$ is a good compromise figure and would produce the following weighting series:

$$\alpha + \alpha(1-\alpha) + \alpha(1-\alpha)^2 + \alpha(1-\alpha)^3 + \alpha(1-\alpha)^4 + \alpha(1-\alpha)^5 \ldots, \text{ etc.}$$
$$0.2 + 0.16 + 0.128 + 0.1024 + 0.08192 + 0.06554 + 0.052429 \ldots, \text{ etc.}$$

which already sums to 0.79 and will clearly sum to one at infinity.

On first examination, a forecast based on Equation [2.2] would appear to be relatively complicated to implement and with an infinite number of demand values hardly able to solve the problem of the amount of data that has to be stored (already referred to as a disadvantage of the moving average). However, it is possible (see Appendix B for a detailed development of this form of the exponentially weighted average) to show that Equation [2.2] can be simplified to the simple statement such that a one-period-ahead forecast f_{t+1} would be of the form:

$$f_{t+1} = u_t = \alpha d_t + (1-\alpha)u_{t-1} \quad [2.3]$$

which is the equivalent of:

$$f_{t+1} = u_t = u_{t-1} + \alpha(d_t - u_{t-1})$$

and since the current forecasting error, e_t, can be defined as the current demand value, d_t, minus the one-period-ahead forecast evaluated last period u_{t-1}, i.e. $e_t = d_t - u_{t-1}$.

It then follows that Equation [2.3] can be simplified to:

$$f_{t+1} = u_t = u_{t-1} + \alpha e_t \quad [2.4]$$

in which form it can more easily be interpreted, for those not too comfortable with simple algebra, as:

New forecast = old forecast + alpha*current forecasting error

Advantages of the simple exponentially weighted average

In contrast to the moving average, the simple exponentially weighted average offers the following advantages:

- it is easy to initialise, since once an estimate for u_{t-1} (the previous period's forecast) is made, forecasting can proceed given that a value for α is also decided since all the unknowns on the right-hand side of Equation [2.3] are then defined;
- it is economical in data storage terms since u_{t-1} embodies all previous data (i.e. $u_{t-1} = \alpha d_{t-1} + \alpha(1 - \alpha)d_{t-2}\ldots$) and hence only the value of u_{t-1} needs to be retained from one period to the next;
- it is capable of having its sensitivity changed at any time by altering the value of α (the exponential smoothing constant) just as long as the value of α is set between zero and one; and
- it does not suddenly cut off in weighting of demand data irrespective of age. Even data 20 periods old is theoretically weighted with $\alpha(1-\alpha)^{20}$ which, although very small irrespective of the value of α, is certainly not zero.

The calculations involved in evaluating the simple exponentially weighted average as a forecasting model based on Equation [2.3], as displayed by the FOREMAN's EWA file, are shown in Fig. 2.1 for a value of $\alpha = 0.2$. The

C.A.L. Module OPERATIONS CONTROL – Forecasting CDL/96

SIMPLE ADAPTIVE DELAYED ADAPT. PRINT VIEWGRAPH GRAPHSAVE PCD QUIT

Simple exponentially weighted average forecasting model

Simple exponentially weighted average forecast

Period........	1	2	3	4	5	6	7	8	9	10	11
A:Current demand	55	50	58	49	86	52	54	49	58	68	75
B:Past forecast	50	51	51	52	51	58	57	56	55	56	58
C:Error	5	-1	7	-3	35	-6	-3	-7	3	12	17
D:Cumul. error	5	4	11	8	43	37	34	27	30	42	59
E:Squared error	25	1	49	9	1225	36	9	49	9	144	289
F:Cum. sq. error	25	26	75	84	1309	1345	1354	1403	1412	1556	1845
G:ALPHA'xError	1.0	-0.2	1.4	-0.6	7.0	-1.2	-0.6	-1.4	0.6	2.4	3.4
H:(1-ALPHA')PSE	0.0	0.8	0.5	1.5	0.7	6.2	4.0	2.7	1.0	1.3	3.0
I:Current SE	1.0	0.6	1.9	0.9	7.7	5.0	3.4	1.3	1.6	3.7	6.4
J:ALPHA'xAbs.Err	1.0	0.2	1.4	0.6	7.0	1.2	0.6	1.4	0.6	2.4	3.4
K:(1-ALPHA')PMAD	4.0	4.0	3.4	3.8	3.5	8.4	7.7	6.6	6.4	5.6	6.4
L:Current MAD	5.0	4.2	4.8	4.4	10.5	9.6	8.3	8.0	7.0	8.0	9.8
M:Standard dev.	6.8	5.3	6.0	5.5	13.2	12.0	10.4	10.0	8.8	10.0	12.3
N:Tracking sign.	0.2	0.1	0.4	0.2	0.7	0.5	0.4	0.2	0.2	0.5	0.6
O:ALPHA (EWC)	0.2	0.2	0.2	0.2	0.2	0.2	0.2	0.2	0.2	0.2	0.2
P:ALPHAxCur.Dem.	11.0	10.0	11.6	9.8	17.2	10.4	10.8	9.8	11.6	13.6	15.0
Q:(1-ALPHA)PFcst	40.0	40.8	40.8	41.6	40.8	46.4	45.6	44.8	44.0	44.8	46.4
R:New forecast	51.0	50.8	52.4	51.4	58.0	56.8	56.4	54.6	55.6	58.4	61.4

File:EWA Esc toggles menu MENU

2.1 The screen display from FOREMAN's EWA file with the calculations of the simple exponentially weighted average forecast with $\alpha = 0.2$ taking place in rows O, P and Q producing a one-period-ahead forecast in row R which is then carried forward to the next period as an integer value in row B.

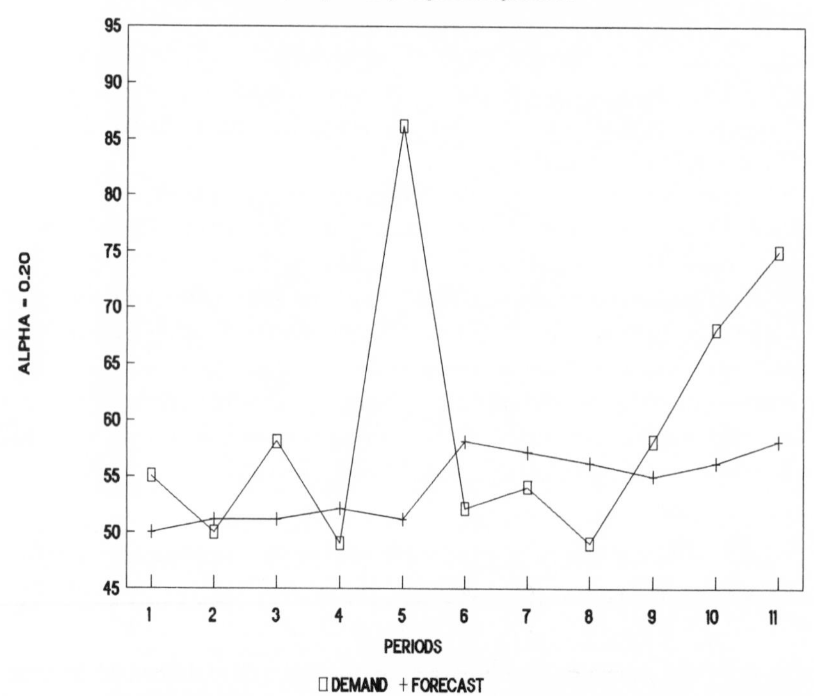

2.2 Response of a simple exponentially weighted average forecast with $\alpha = 0.2$ using FOREMAN's EWA file with accompanying demonstration data.

resulting graph of demand values and associated one-period-ahead forecasts associated with these calculations is shown as Fig. 2.2.

Examining the response of the simple exponentially weighted average forecast based on $\alpha = 0.2$ as shown in Fig. 2.2, it should be noted that:

- for the first four periods, during which the demand is fairly stable, as would be expected the forecast falls within the demand values;
- the impulse (one individual, large demand value) occurring in period 5 is not anticipated;
- as a result of reacting one period late to the impulse in period 5, the forecast in period 6 is higher than might be expected by an amount equal to one-fifth (since $\alpha = 0.2$) of the forecasting error produced in period 5; and
- in periods 8 to 11, where clearly a growth of demand has occurred, the forecast has failed to react and lags behind the growth. This indicates

Task 2.1. Using FOREMAN's EWA file establish the best value of α for the SIMPLE exponential smoothing forecasting model used for analysing the data presented in that file. Use the minimum sum of squared errors as the criterion for choosing the value of α and then explain why the chosen value is the best.

why the simple version of the exponentially weighted average is not suitable for demand data which exhibits growth characteristics.

For the simple exponentially weighted average, when the value of (alpha) α is high, a good response to an upward change can be anticipated. However, with a high value of α a single high demand value (impulse) can cause an over-reaction one period late (i.e. the forecast is so sensitive it can over-react to any 'noise' in the demand process). Conversely, when the value of α is low although the effect of an impulse will be ignored, the response to an upward change will be poor.

Statistics for establishing the validity and accuracy of forecasting models

In trying to establish which forecast is best in any particular situation, it is necessary to have statistical information available, particularly with regard to the size of the forecasting errors. The two most used statistics for selecting the suitability of forecasting models are now described in detail.

The Mean Squared Error

The Mean Squared Error (MSE) is the average of the squared forecasting errors. As such it is often the statistic used to ascertain the best forecasting model, it being assumed that the model with the minimum MSE will be best where:

$$\text{MSE} = \frac{1}{n}\sum_{t=1}^{n} e^2 \qquad [2.5]$$

NOTE: that within FOREMAN's EWA file the sum of squared errors $\sum_{t=1}^{n} e^2$ rather than the MSE $\frac{1}{n}\sum_{t=1}^{n} e^2$ is used since the formulation of this measure can be displayed directly within the numerical display.

The Mean Absolute Percentage Error

The Mean Absolute Percentage Error (MAPE) is one of the most commonly used statistics in all types of forecasting. It gives an indication of the average size of forecasting error expressed as a percentage of the relevant observed value, irrespective of whether that forecasting error is positive (where the forecast under-estimates) or negative (where the forecast over-estimates).

In computational terms, if the forecasting error at time, t, (i.e. e_t) is defined as the demand, d_t, minus the forecast, f_t $(=u_{t-1})$, such that $e_t=(d_t-u_{t-1})$, it then follows that the MAPE is defined as:

$$\text{MAPE} = \frac{100}{n}\sum_{t=1}^{n}\frac{|e_t|}{d_t} \qquad [2.6]$$

where:

- the | | symbols either side of e_t represent absolute values and, hence, $|e_t|$ is considered positive irrespective of whether e_t is positive or negative; and
- n represents the number of observations involved.

NOTE: Where zero demand values exist, these must be excluded from the MAPE calculation simply because the result of dividing by zero produces a value of infinity

Because the MAPE measures the average relative size of the absolute forecasting error as a percentage of the corresponding demand value, in practice a value of less than 10% would be regarded as a very good fit and providing potentially very good forecasts. As a relative measure, its interpretation can be broadly categorised as shown in Table 2.2.

Both the MSE and the MAPE are calculated and displayed in the forecasting analysis carried out by FOREMAN's SIMPLE file as shown in Fig. 2.3. The screen layout for both the SIMPLE and the TREND files (the latter's function is described in Chapter 4) produces the results for three forecasting models and their associated statistics. In the particular case of the simple exponentially weighted average forecast with $\alpha=0.1$, for the DEMONSTRATION data STATIONARY these can be seen to be MSE = 109 (i.e. 1.09E+2 = 1.09×10^2) and MAPE = 17.88%.

Table 2.2 Values of the MAPE and related forecast potential

MAPE <10% – forecasts potentially very good
MAPE <20% – forecasts potentially good
MAPE <30% – forecasts potentially reasonable
MAPE >30% – forecasts potentially inaccurate

The response of the simple exponentially weighted average forecasting model to different demand characteristics

As has been indicated earlier, the sensitivity of an exponentially weighted average clearly depends on the value of (alpha) α – the exponential smoothing constant (EWC) – which can only vary between zero and one. At the two extreme values of α the following occur:

- if $\alpha = 0$, then from Equation [2.3] $f_{t+1} = u_{t-1}$ and the one-period-ahead forecast remains fixed at the value of the previous forecast, i.e. the forecast is totally insensitive to changes in the demand pattern; or
- if $\alpha = 1$, then $f_{t+1} = d_t$ and the one-period-ahead forecast is equal to the most recent period's demand value, i.e. the forecast is extremely sensitive to changes in the demand pattern and can, therefore, over-react to relatively small changes.

Values of α of either 0.1 or 0.2 are useful compromise figures which are often used in practice and, as can be seen from Table 2.3, equate to an equivalent moving average of $n = 19$ or $n = 9$ periods respectively.

The analysis shown in Fig. 2.3 is for a stationary demand series which can be loaded as a DEMONSTRATION set of demand values called STATIONARY. Of the three forecasting models shown, the third column

```
C.A.L. Module              OPERATIONS CONTROL - Forecasting                    CDL/96
SIMPLE EWA  ADAPTIVE.RES.RATE   DEL.ADAP.RES.RATE   CONTINUE
Plot simple ewa

          File: DEMODATA, Range: STATIONVALUES:     29  HORIZON:       12
   ========================================================================
    PERIOD       DATA   SIMPLE   ADAPTIVE  DEL ADAP
      t           y      EWA     RES RATE  RES RATE
   ========================================================================
         1        50    50.00     50.00     50.00       ALPHA =       0.1
         2        55    50.00     50.00     50.00
         3        45    50.50     51.19     50.00      MEAN SQUARED ERROR
         4        48    49.95     50.60     48.81      ------------------
         5        56    49.76     50.06     48.77      SIMPLE   1.09E+02
         6        47    50.38     50.70     49.41      ADAPTIVE 1.21E+02
         7        59    50.04     50.41     48.75      DEL ADAP 1.60E+02
         8        46    50.94     52.78     49.96
         9        54    50.44     52.53     48.13      MEAN ABSOLUTE
        10        67    50.80     52.57     49.48            M.A.P.E.
        11        34    52.42     59.05     56.59      ------------------
        12        66    50.58     52.71     41.26      SIMPLE      17.88%
        13        36    52.12     53.40     42.13      ADAPTIVE    18.67%
        14        68    50.51     49.09     40.01      DEL ADAP    19.09%
        15        37    52.26     50.98     45.83
  File:SIMPLE                  Esc toggles menu                              MENU
```

2.3 The screen display from FOREMAN's SIMPLE file with forecast values based on a simple exponentially weighted average model with $\alpha = 0.1$ for the DEMONSTRATION demand series STATIONARY together with the MSE and MAPE statistics.

Table 2.3 Table of equivalent values of (alpha) α used to form an exponential weighted average and n the number of periods included within a moving average of the same 'average age of data'

(alpha) α	n
0.05	39
0.1	19
0.154	12
0.2	9
0.3	6

from the left (i.e. the left-hand of the three forecasting models) reveals the forecasts based on the simple exponential weighted average forecast with $\alpha = 0.1$. Figure 2.4 shows the corresponding graphical display which demonstrates for this STATIONARY series that the simple exponentially weighted average forecast successfully tracks the central tendency and produces a sensible set of forecasts up to twelve periods ahead.

Although the stationary demand situation depicted in Fig. 2.4 is ideally

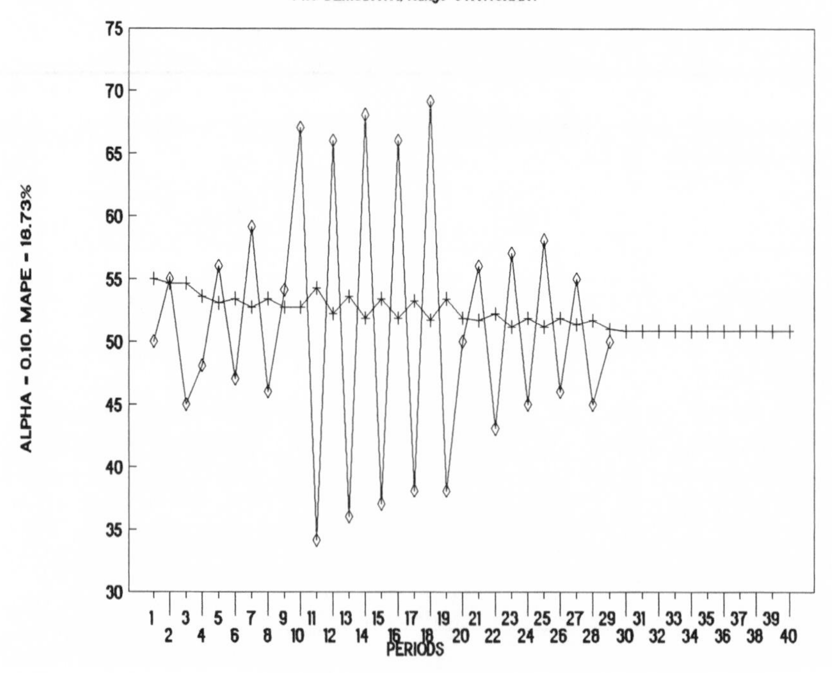

2.4 Response of a simple exponentially weighted average forecast with $\alpha = 0.1$ using FOREMAN's SIMPLE file to analyse the DEMONSTRATION demand series STATIONARY

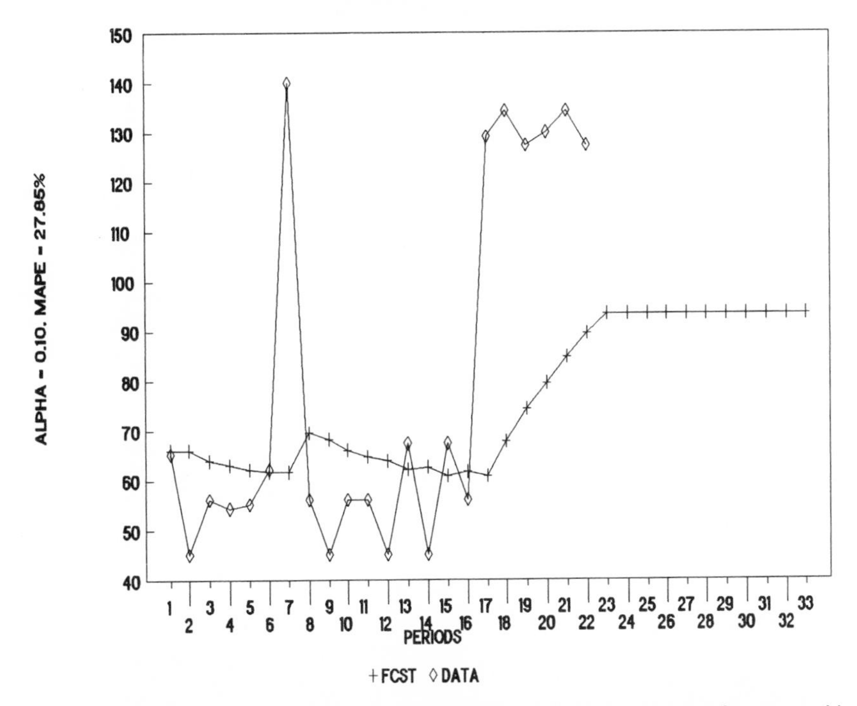

2.5 Response of a simple exponentially weighted average forecast with $\alpha = 0.1$ using FOREMAN's SIMPLE file to analyse the DEMONSTRATION demand series IMPSTEP.

suited to the simple exponentially weighted average forecasting model, clearly this forecasting model's response to other demand characteristics needs to be examined.

Examining Fig. 2.5, which depicts the response of a simple exponentially weighted average forecast based on $\alpha = 0.1$ using FOREMAN's SIMPLE files with the DEMONSTRATION demand series IMPSTEP demonstrates that:

- when a single, high demand value (impulse) occurs, a simple exponentially weighted average forecast reacts one period late and then responds to the impulse by an amount equal to α of the height of the impulse; and
- when a simple exponentially weighted average forecast encounters a step change (i.e. a higher than average demand value which is sustained by further high values) it responds progressively, for each passing time

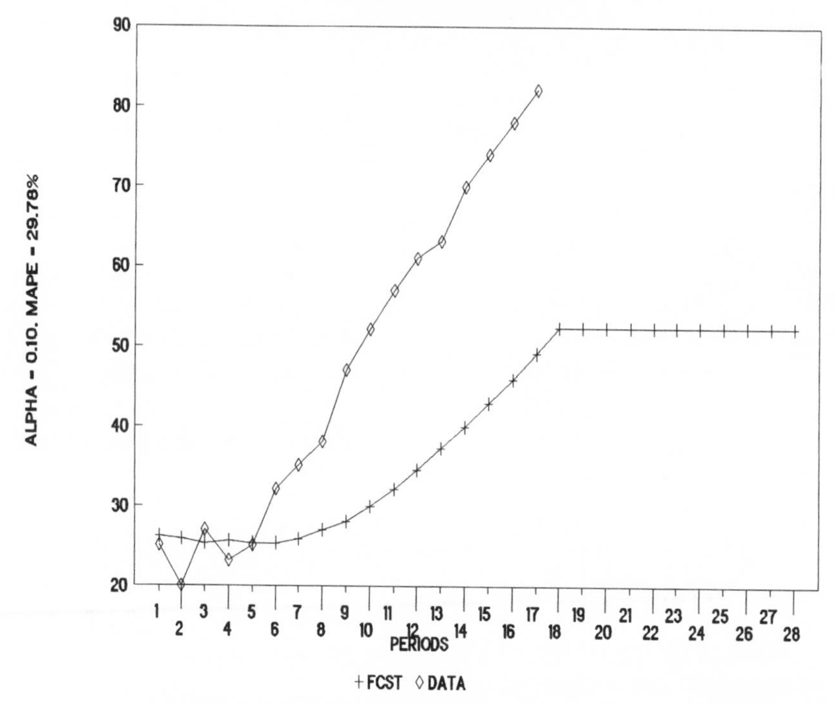

2.6 Response of a simple exponentially weighted average forecast with $\alpha = 0.1$ using FOREMAN's SIMPLE file to analyse a short version of the DEMONSTRATION demand series POSIRAMP.

period, by a proportion α of the remaining size of that change. With a value of $\alpha = 0.1$, this asymptotic form of response will be very slow.

Examination of Fig. 2.6 indicates why the simple exponentially weighted average forecasting model is not suitable for demand situations which are subject to growth. Here, using FOREMAN's SIMPLE file with a DEMONSTRATION demand series POSIRAMP, an exponentially weighted average forecast based on $\alpha = 0.1$ demonstrates that:

- the forecast lags behind the growth and, if the growth were to continue, would never catch up; and
- because no growth is assumed, forecast ahead are fixed at the value of the forecast one period ahead of the last demand value.

Because of the inability of the simple exponentially weighted average to cope with growth, more sophisticated models based on the exponentially

---

Task 2.2. Using FOREMAN's DATAHELP file create a data file of stationary demand figures. Then use the file SIMPLE to analyse that data file and find the best value of the exponential weighting constant (alpha) α for the simple exponential weighted average forecasting model. Use the minimum mean squared error as the criterion for choosing the value of α.

---

weighted average concept are required when growth in demand is significant.

Monitoring forecasting systems

Within any forecasting system it is necessary to monitor the accuracy of the forecasts being produced and to correct manually those forecasts which go out of control due to changes in the demand pattern. In this section, the monitoring of short-term forecasts is discussed with particular emphasis placed on those situations where many stocked item forecasts are being produced to establish inventory control parameters.

Most forecasting systems which involve many items operate on the basis that if there is no evidence to the contrary then it is assumed that the forecast is in control, i.e. there have been no significant changes in the demand pattern to make current forecasts invalid. For such a policy of management by exception to operate successfully, clearly an effective monitoring system is essential.

Although several different approaches have been taken with regard to monitoring forecasts, Trigg's[9] proposal for a tracking signal (known in the US as the smoothed error tracking signal) has become an essential part of the majority of comprehensive short-term forecasting systems.

The Trigg or smoothed error tracking signal

The tracking signal proposed originally by Trigg is based on the fact that if forecasting errors, e_t, are defined as demand minus forecast (i.e. $e_t = d_t - u_{t-1}$) then the current smoothed error, $\bar{e}_t$, is defined as the exponentially weighted average of the forecasting errors, e_t, and is produced by:

$$\bar{e}_t = \alpha' e_t + (1 - \alpha')\bar{e}_{t-1} \qquad [2.7]$$

where $\bar{e}_{t-1}$ is the value of the smoothed error for the previous (i.e. immediate past) time period.

The current value of mean absolute deviation (MAD) is then defined as the exponentially weighed average of the absolute forecasting errors, $|e_t|$, using the formula:

$$\mathrm{MAD}_t = \alpha'|e_t| + (1 - \alpha')\mathrm{MAD}_{t-1} \quad [2.8]$$

where the absolute value signs | | indicate that all errors, e_t, are treated as positive irrespective of their actual polarity, and where MAD_{t-1} is the value of the mean absolute deviation for the previous time period.

In both equations [2.7] and [2.8] the parameter (alpha') α' is an exponential weighting constant (EWC) whose value must be between zero and one. Although the value of α' may be the same as the value of α used in producing the exponentially weighted average forecast, by convention for monitoring applications α' is usually set at a fixed value of $\alpha' = 0.2$.

Having defined the smoothed error, $\bar{e}_t$, and the mean absolute deviation, MAD_t, the tracking signal, T_t, is then defined as the ratio of the smoothed error to the mean absolute deviation, hence:

$$T_t = \bar{e}_t \,/\, \mathrm{MAD}_t \quad [2.9]$$

Given that the value of α' used to produce both $\bar{e}_t$ and MAD_t are the same and set at 0.2, then in practice, irrespective of the data involved, the value of the tracking signal can only vary between +1 and −1 and can be interpreted statistically as follows:

- if the value of the tracking signal exceeds 0.7, the user can be 95% confident in the hypothesis that the accompanying forecast is out of control due to an untypically high set of demand values for which there should be an identifiable, external cause. This could, for instance, be due to the failure of a competitor supplier to meet commitments; and
- if the value of the tracking signal is less than −0.7, the user can be 95% confident in the hypothesis that the accompanying forecast is out of control due to an untypically low set of demand values for which there should be an identifiable, external cause. This could, for instance, be due to the loss of a major customer or a general loss of custom due to financial or legislative changes.

The calculations involved in producing the tracking signal are demonstrated in FOREMAN's EWA file, which is shown as Fig. 2.7, as follows:

- The current smoothed error *Current SE* (i.e. $\bar{e}_t$) is calculated in row I as the sum of ALPHA' multiplied by the current error *ALPHA' × Error* (i.e. $\alpha' e_t$) in row G and one minus ALPHA' multiplied by the past smoothed error *(1 − ALPHA')PSE* (i.e. $(1 - \alpha')\bar{e}_{t-1}$) in row H. Note the initial value of the past smoothed error PSE is by convention set equal to zero (see row H period 1).

C.A.L. Module OPERATIONS CONTROL – Forecasting CDL/96

SIMPLE ADAPTIVE DELAYED ADAPT. PRINT VIEWGRAPH GRAPHSAVE LCD QUIT

Simple exponentially weighted average forecasting model

Simple exponentially weighted average forecast

Period........	1	2	3	4	5	6	7	8	9	10	11
A Current demand	55	50	58	49	86	52	54	49	58	68	75
B Past forecast	50	51	51	52	51	58	57	56	55	56	58
C Error	5	-1	7	-3	35	-6	-3	-7	3	12	17
D Cumul. error	5	4	11	8	43	37	34	27	30	42	59
E Squared error	25	1	49	9	1225	36	9	49	9	144	289
F Cum. sq. error	25	26	75	84	1309	1345	1354	1403	1412	1556	1845
G ALPHA'xError	1.0	-0.2	1.4	-0.6	7.0	-1.2	-0.6	-1.4	0.6	2.4	3.4
H (1-ALPHA')PSE	0.0	0.8	0.5	1.5	0.7	6.2	4.0	2.7	1.0	1.3	3.0
I Current SE	1.0	0.6	1.9	0.9	7.7	5.0	3.4	1.3	1.6	3.7	6.4
J ALPHA'xAbs.Err	1.0	0.2	1.4	0.6	7.0	1.2	0.6	1.4	0.6	2.4	3.4
K (1-ALPHA')PMAD	4.0	4.0	3.4	3.8	3.5	8.4	7.7	6.6	6.4	5.6	6.4
L Current MAD	5.0	4.2	4.8	4.4	10.5	9.6	8.3	8.0	7.0	8.0	9.8
M Standard dev.	6.3	5.3	6.0	5.5	13.2	12.0	10.4	10.0	8.8	10.0	12.3
N Tracking sign.	0.2	0.1	0.4	0.2	0.7	0.5	0.4	0.2	0.2	0.5	0.6
O ALPHA (EWC)	0.2	0.2	0.2	0.2	0.2	0.2	0.2	0.2	0.2	0.2	0.2
P ALPHAxCur.Dem.	11.0	10.0	11.6	9.8	17.2	10.4	10.8	9.8	11.6	13.6	15.0
Q (1-ALPHA)PFcst	40.0	40.8	40.8	41.6	40.8	46.4	45.6	44.8	44.0	44.8	46.4
R New forecast	51.0	50.8	52.4	51.4	58.0	56.8	56.4	54.6	55.6	58.4	61.4

File:EWA Esc toggles menu MENU

2.7 FOREMAN's EWA file's screen display highlighting the calculations required to produce the smoothed error tracking signal when the forecast is a simple exponentially weighted average with $\alpha = 0.2$.

- The current mean absolute deviation *Current MAD* (i.e. MAD_t) is calculated in row L as the sum of ALPHA′ multiplied by the current absolute error *ALPHA′ × Abs.Err.* (i.e. $\alpha'|e_t|$) in row J and one minus ALPHA′ multiplied by the past mean absolute deviation *(1 – ALPHA′)PMAD* (i.e. $(1-\alpha')MAD_{t-1}$) in row K. The initial value of the past mean absolute deviation PMAD is by convention set equal to one-tenth of the initial forecast estimate u_{t-1} (in this case $0.1*50 = 5.0$ hence in row K period 1 the value $0.8*5.0 = 4.0$). This convention avoids the illogical situation of the tracking signal starting at either -1 or $+1$.
- The current value of the tracking signal T_t *Tracking sign* is calculated in row N as the ratio of the current smoothed error *Current SE* (i.e. $\bar{e}_t$) in row I and the current mean absolute deviation *Current MAD* (i.e. MAD_t) in row L.

Further examination of Fig. 2.7, which is the screen display of FOREMAN's EWA file, shows that the value of the tracking signal equals 0.7 in period 3 when the forecast being monitored is a simple exponentially weighted average based on a value of $\alpha = 0.2$. This high value confirms the detection of a significant, positive, single period impulse of 86 occurring in that period. The occurrence of this impulse can also be verified by referring back to Fig. 2.2 (page 26).

Task 2.3. Using FOREMAN's EWA file:

(1) Noting that when a simple exponentially weighted average forecasting model based on a value of α=0.2 is used, the forecast does go out of control due to a single period impulse in period 3, consider a practical range of α from 0.1 to 0.5 and confirm that the tracking signal's ability to detect this out of control situation is unaffected by changes in the value of α used to produce the forecast.

(2) What other change in the demand pattern is also detected as an 'out of control' situation?

Implementing a monitoring system based on the Trigg or smoothed error tracking signal

To operate a monitoring system based on the principle of management by exception, a comprehensive method of implementing the smoothed error tracking signal would be to:

- calculate the value of the tracking signal for all items;
- exclude from consideration those items for which the tracking signal had already indicated an out of control situation;
- of those items that remain, list those items for which the absolute value of the tracking signal exceeds a value of 0.7 in descending order of the absolute value of the tracking signal; and
- investigate the top N items in the resulting list as items for which the forecasting system is now indicating it can no longer produce sensible forecasts, where N is the number of items for which there are sufficient resources to investigate the reasons for the out of control situation being created and for which manual corrections need to be made.

Adaptive forecasting

Adaptive forecasting describes a family of forecasting models whose parameters 'adapt' to the volatility of the demand data being analysed. In particular, with regard to exponential weighted average, it could be argued that a high value of α would be desirable to produce a sensitive forecast when the data being analysed were relatively volatile whereas a low value of α producing a less sensitive forecast would be more appropriate when the data were relatively stable.

The adaptive response rate forecast

One way that the value of the exponential weighting constant could be made to respond to the volatility of the demand data, as suggested by Trigg and Leach,[10] would be to set the value of α within an exponential weighted average forecasting model equal to the absolute value (or modulus) of the tracking signal (i.e. $|T_{t.}|$). Such a proposal is certainly feasible since because the tracking signal value can only vary between −1 and +1, the absolute value must clearly only vary between 0 and +1 and, therefore, meets the criterion required of α to form a true average whose weights must sum to one at infinity (see page 24).

Hence with $\alpha = |T_t|$, the formula for an adaptive form of the simple exponentially weighted average would be:

$$\tilde{u}_t = |T_t|d_t + (1 - |T_t|)\tilde{u}_{t-1} \qquad [2.10]$$

The calculations involved in this adaptive response rate forecasting model are highlighted in Fig. 2.8 which shows FOREMAN's EWA file's screen display when the ADAPTIVE option is chosen. Note in row O that ALPHA (EWC) (i.e. α) is set equal to the absolute value of the tracking signal value calculated in row N.

The improvement in the response of the adaptive response rate forecast compared with a simple exponentially weighted average to a genuine step change in demand can be seen in Fig. 2.9. However, from Fig. 2.9 it is also

C.A.L. Module OPERATIONS CONTROL – Forecasting CDL/96
SIMPLE ADAPTIVE DELAYED ADAPT. PRINT VIEWGRAPH GRAPHSAVE LCD QUIT
Adaptive response rate forecasting model

Adaptive response rate forecast

Period........	1	2	3	4	5	6	7	8	9	10	11
A Current demand	55	50	58	49	86	52	54	49	58	68	75
B Past forecast	50	51	51	54	53	75	73	69	61	60	62
C Error	5	-1	7	-5	33	-23	-19	-20	-3	8	13
D Cumul. error	5	4	11	6	39	16	-3	-23	-26	-18	-5
E Squared error	25	1	49	25	1089	529	361	400	9	64	169
F Cum. sq. error	25	26	75	100	1189	1718	2079	2479	2488	2552	2721
G ALPHA'xError	1.0	-0.2	1.4	-1.0	6.6	-4.6	-3.8	-4.0	-0.6	1.6	2.6
H (1-ALPHA')PSE	0.0	0.8	0.5	1.5	0.4	5.6	0.8	-2.4	-5.1	-4.6	-2.4
I Current SE	1.0	0.6	1.9	0.5	7.0	1.0	-3.0	-6.4	-5.7	-3.0	0.2
J ALPHA'xAbs.Err	1.0	0.2	1.4	1.0	6.6	4.6	3.8	4.0	0.6	1.6	2.6
K (1-ALPHA')PMAD	4.0	4.0	3.4	3.8	3.8	8.4	10.4	11.3	12.3	10.3	9.5
L Current MAD	5.0	4.2	4.8	4.8	10.4	13.0	14.2	15.3	12.9	11.9	12.1
M Standard dev.	6.3	5.3	6.0	6.0	13.1	16.2	17.7	19.2	16.1	14.9	15.1
N Tracking sign.	0.2	0.1	0.4	0.1	0.7	0.1	-0.2	-0.4	-0.4	-0.3	0.0
O ALPHA (EWC)	0.2	0.1	0.4	0.1	0.7	0.1	0.2	0.4	0.4	0.3	0.0
P ALPHAxCur.Dem.	11.0	7.1	22.9	5.1	57.7	4.0	11.4	20.4	25.8	17.0	1.4
Q (1-ALPHA)PFcst	40.0	43.7	30.9	48.3	17.5	69.2	57.6	40.2	33.9	45.0	60.9
R New forecast	51.0	50.9	53.8	53.5	75.1	73.2	69.0	60.7	59.7	62.0	62.2

File:EWA Esc toggles menu MENU

2.8 FOREMAN's EWA file's screen display showing the calculations required to evaluate the adaptive response rate forecast. Note that the absolute value of the tracking signal becomes the value of the exponential weighting constant used to produce the next forecast, i.e. $\alpha = |T_t|$

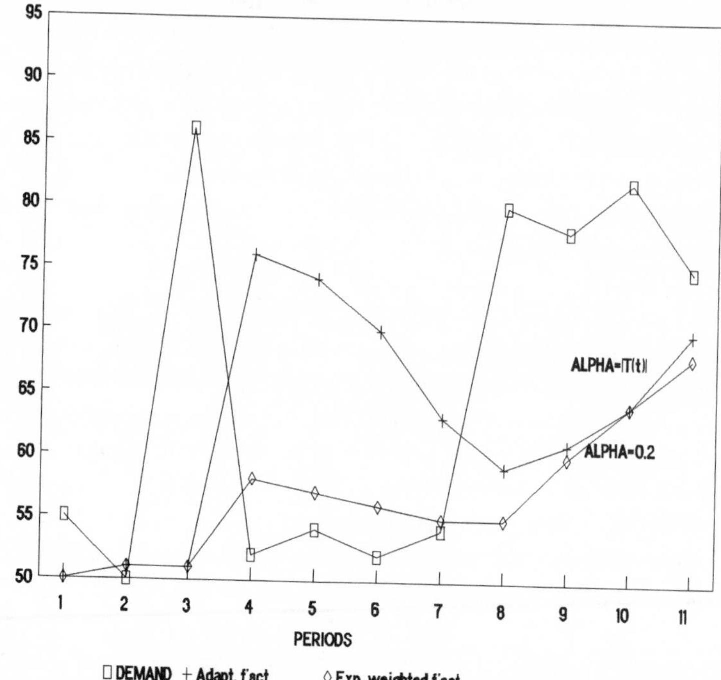

2.9 Response of an adaptive response rate forecast with $\alpha = |T_t|$ to a single period impulse followed by a step change as generated by FOREMAN's EWA file. The author has superimposed a simple exponentially weighted average forecasting model with $\alpha = 0.2$ for comparison.

clear that when the adaptive response rate forecast encounters a single period impulse, it over-reacts one period late. This is because, like all other forecasts, the adaptive response rate forecast cannot foresee a sudden impulse and since it generates a high value of the tracking signal, T_t, at the time of the impulse it subsequently produces a forecast following the impulse as a combination of the value of the previous forecast plus a significant proportion (a high value of $\alpha = |T_t|$) of the forecasting error, which in effect is the height of the impulse. Hence the inevitable over-reaction.

The delayed adaptive response rate forecast

Because of its over-reaction to a single period impulse, the adaptive response rate forecast has largely been replaced in practice by the delayed adaptive response rate forecast; an alternative version proposed by Shone.[8] The delayed version is of the form:

$$\tilde{u}_t = |T_{t-1}|d_t + (1 - |T_{t-1}|)\tilde{u}_{t-1} \qquad [2.11]$$

As can be seen from Equation 2.11, for this delayed version of the adaptive response rate forecast, the exponential weighting constant, α, is set equal to the absolute value of the one period delayed value of the tracking signal rather than the current version $|T_t|$ (i.e. $\alpha = |T_{t-1}|$). This slight modification cancels out the one period delayed over-reaction to a single period impulse without impairing the response to genuine step change. The necessary calculations involved in this delayed adaptive response rate forecast are highlighted in Fig. 2.10, which shows FOREMAN's EWA file's screen display when the DELAYED option has been chosen. Note in row O that ALPHA (EWC) (i.e. α) is set equal to the absolute value of the one period delayed tracking signal, $|T_{t-1}|$, calculated in row N. By convention the initial value of $\alpha = |T_{t-1}|$ is set at 0.2 as evidenced in row N period 1.

The improved response of the delayed adaptive response rate forecast in not responding one period late to a single period impulse but still responding well to a step change can be seen in Fig. 2.11.

C.A.L. Module OPERATIONS CONTROL – Forecasting CDL/96

SIMPLE ADAPTIVE DELAYED ADAPT. PRINT VIEWGRAPH GRAPHSAVE LCD QUIT

Delayed adaptive response rate forecasting model

Delayed adaptive response rate forecast

Period........	1	2	3	4	5	6	7	8	9	10	11
A Current demand	55	50	58	49	86	52	54	49	58	68	75
B Past forecast	50	51	51	52	51	58	54	54	51	53	60
C Error	5	-1	7	-3	35	-6	0	-5	7	15	15
D Cumul. error	5	4	11	8	43	37	37	32	39	54	69
E Squared error	25	1	49	9	1225	36	0	25	49	225	225
F Cum. sq. error	25	26	75	84	1309	1345	1345	1370	1419	1644	1869
G ALPHA'xError	1.0	-0.2	1.4	-0.6	7.0	-1.2	0.0	-1.0	1.4	3.0	3.0
H (1-ALPHA')PSE	0.0	0.8	0.5	1.5	0.7	6.2	4.0	3.2	1.7	2.5	4.4
I Current SE	1.0	0.6	1.9	0.9	7.7	5.0	4.0	2.2	3.1	5.5	7.4
J ALPHA'xAbs.Err	1.0	0.2	1.4	0.6	7.0	1.2	0.0	1.0	1.4	3.0	3.0
K (1-ALPHA')PMAD	4.0	4.0	3.4	3.8	3.5	8.4	7.7	6.2	5.7	5.7	7.0
L Current MAD	5.0	4.2	4.8	4.4	10.5	9.6	7.7	7.2	7.1	8.7	10.0
M Standard dev.	6.3	5.3	6.0	5.5	13.2	12.0	9.6	8.9	8.9	10.9	12.5
N Tracking sign.	0.2	0.1	0.4	0.2	0.7	0.5	0.5	0.3	0.4	0.6	0.7
O ALPHA (EWC)	0.2	0.2	0.1	0.4	0.2	0.7	0.5	0.5	0.3	0.4	0.6
P ALPHAxCur.Dem.	11.0	10.0	8.3	19.4	17.6	38.2	27.9	25.4	17.7	30.0	47.6
Q (1-ALPHA)PFcst	40.0	40.8	43.7	31.5	40.5	15.4	26.1	26.1	35.4	29.6	21.9
R New forecast	51.0	50.8	52.0	50.8	58.2	53.6	54.0	51.4	53.1	59.6	69.5

File:EWA Esc toggles menu MENU

2.10 FOREMAN's EWA file's screen display showing the calculations required to evaluate the delayed adaptive response rate forecast. Note that the one period delayed absolute value of the tracking signal becomes the value of the exponential weighting constant used to produce the next forecast, i.e. $\alpha = |T_{t-1}|$.

DEMAND ANALYSIS - FORECASTING SCHEDULE

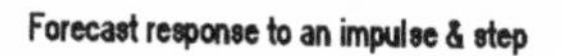

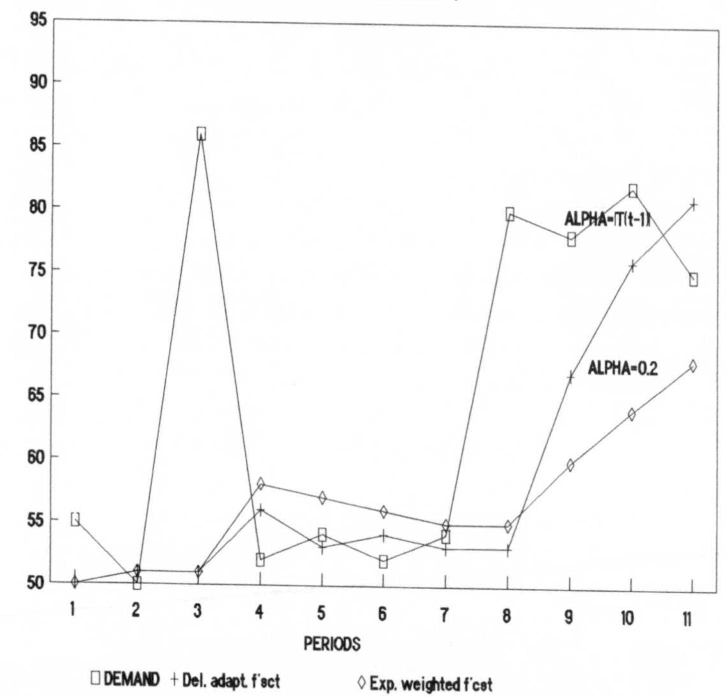

2.11 Response of a delayed adaptive response rate forecast with $\alpha = |T_{t-1}|$ to a single period impulse followed by a step change as generated by FOREMAN's EWA file. The author has superimposed a simple exponentially weighted average forecasting model with $\alpha = 0.2$ for comparison.

Task 2.4. Using FOREMAN's SIMPLE file establish the best forecast for the IMPSTEP demonstration data between:

(1) the simple exponentially weighted average with the best value of α

(2) the adaptive response rate forecast; and

(3) the delayed adaptive response rate forecast.

Justify your reasoning.

Conclusion

The simple exponentially weighted average represents an ideal forecasting model for producing relatively short-term forecasts for inventory control systems when demand is stationary. Its advantages are that:

- it is easy to start up for new products;

- it is economical in data storage terms;
- it is easily modified in terms of sensitivity of response; and
- it is generally robust.

However, this simple version of the exponentially weighted average has limitations when on the one hand trying to ignore a single period impulse but on the other hand responding to a genuine step change (as shown in Fig. 2.5). It is also totally unsuitable for forecasting when the demand series exhibits growth characteristics (as is shown in Fig. 2.6).

Files from OPSCON's package FOREMAN associated with this chapter

EWA file – a simple, single level menu driven file which displays the arithmetic calculations involved in developing:

- the simple exponentially weighted average forecasting model (with an option for specifying the value of α);
- the adaptive response rate forecasting model; and
- the delayed adaptive response rate forecasting model.

There is only one set of data provided and it is possible to view the response of these three forecasting models to this set of data.

SIMPLE file – a multi-level menu driven file which allows the user to load either a file, or named range within a file of any suitable demand data or three demonstration sets of demand data. The following forecasts are established after an initial forecast value is supplied, namely:

- the simple exponentially weighted average forecasting model (with an option for specifying the value of α);
- the adaptive response rate forecasting model; and
- the delayed adaptive response rate forecasting model.

The MSE and MAPE statistics are provided and options are available to print tables of results or save graphical PIC files of the forecasts' responses.

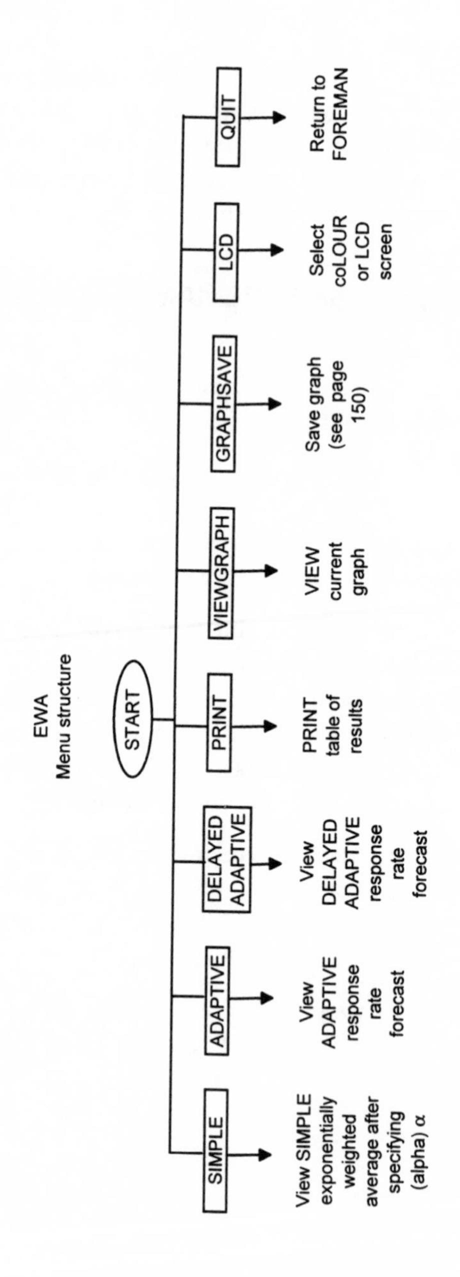
EWA
Menu structure
START
SIMPLE
View SIMPLE exponentially weighted average after specifying (alpha) α
ADAPTIVE
View ADAPTIVE response rate forecast
DELAYED ADAPTIVE
View DELAYED ADAPTIVE response rate forecast
PRINT
PRINT table of results
VIEWGRAPH
VIEW current graph
GRAPHSAVE
Save graph (see page 150)
LCD
Select coLOUR or LCD screen
QUIT
Return to FOREMAN

SIMPLE
Menu structure

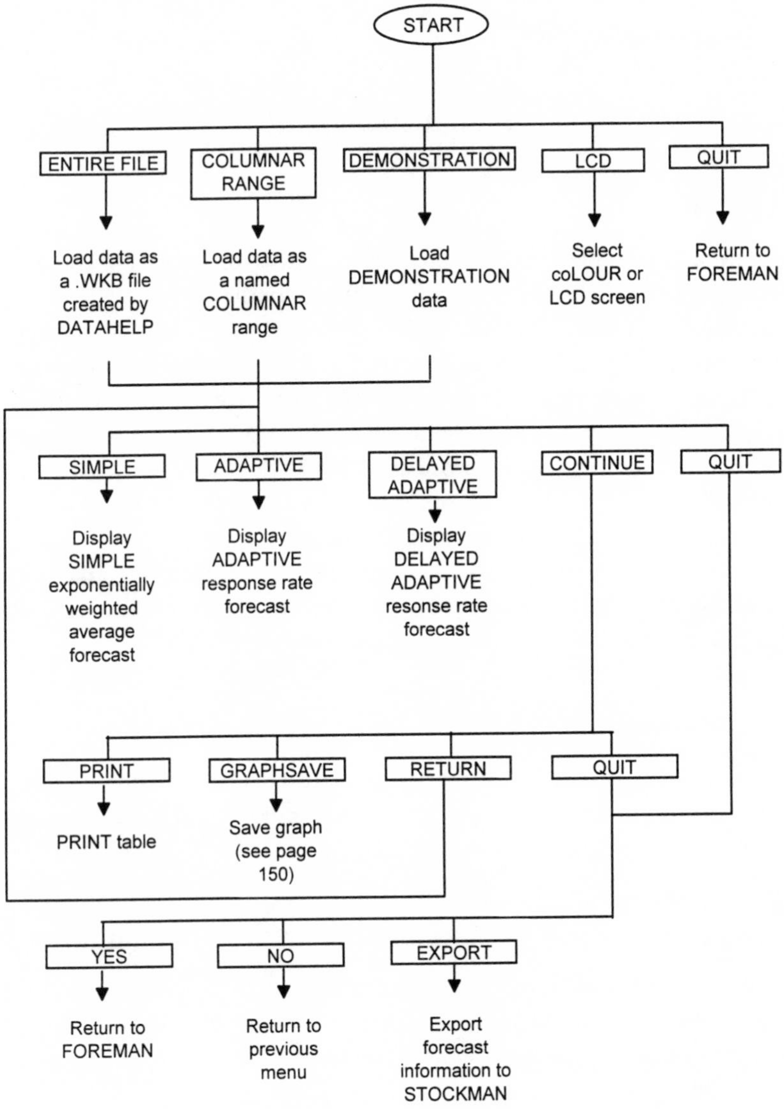

3

Short-term forecasting for growth and seasonality

Introduction

Demand situations which are characterised by either or both growth and seasonal influences represent a more complex forecasting problem than when demand is purely stationary (i.e. no growth or seasonality is assumed). Most forecasting models consider that a stationary element always underpins any demand situation, but that over-and-above that stationary element growth and seasonality could exist. This 'decomposition' approach to forecasting therefore attempts to develop a model which estimates the parameters defining the stationary, growth and seasonality elements assumed to describe the underlying generating process producing the individual demand values. If subsequent to a forecasting analysis the latter two parameters turn out to be insignificant, the forecast can then revert to being based on an assumption of stationary demand.

Growth forecasting models

As was discussed in the previous chapter, the simple exponentially weighted average is not a suitable forecasting model if the data being analysed is subject to growth (see page 32) since it:

- lags behind the data; and
- produces forecasts ahead of the known data which are fixed in value.

Because of these problems, modifications to forecasting models based on the simple exponentially weighted average model are necessary to cope with growth situations where demand increases (or decreases) over a period of time.

A variety of forecasting models have been developed to cope with demand data which exhibit growth. In this chapter Brown's[1] double smoothed model is considered together with Holt's[3] model which is also incorporated in the Holt Winters'[12] seasonal forecasting model.

Brown's double smoothed model

Of the many forecasting models which have been proposed to cope with growth situations, Brown's[1] double smoothed exponentially weighted average forecasting model for growth has the advantage that only one parameter (alpha) α is involved. This makes searching for the optimal forecasting model much simpler than if two or more parameters were involved which would require a two dimensional search procedure. Brown's double smoothed forecasting model is based on the assumption that a double smoothed exponentially weighted average (EWA) of the form:

$$\bar{u}_t = \alpha u_t + (1-\alpha)\,\bar{u}_{t-1} \qquad [3.1]$$

lags behind the simple exponentially weighted average u_t by the same amount as u_t lags behind d_t, the assumed growing demand pattern. As shown previously (see page 24) u_t is defined as:

$$u_t = \alpha d_t + (1-\alpha)u_{t-1} \qquad [3.2]$$

The suggested relationship that u_t lags d_t by the same amount that $\bar{u}_t$ lags behind u_t in the steady state can be confirmed by examining Fig. 3.1.

The double smoothed EWA forecasting model is based on the assumption that:

- the demand data are drawn from a population with a stationary element, μ, and a growth factor, λ, and is of the general form $d_t = \mu + \lambda t + \varepsilon$ where ε represents random errors with zero mean;
- in the steady state that the lag of u_t behind d_t is equal to $\frac{\lambda(1-\alpha)}{\alpha}$; and
- this lag is also equal to the lag of $\bar{u}_t$ behind u_t i.e. $(u_t - \bar{u}_t)$

Based on the assumptions of equal lags made above, it follows that:

$$\frac{\lambda(1-\alpha)}{\alpha} = (u_t - \bar{u}_t)$$

which produces an estimate b_t for the growth factor λ of:

$$b_t = \frac{\alpha(u_t - \bar{u}_t)}{1-\alpha}$$

To overcome the problem that both u_t and $\bar{u}_t$ lag behind the assumed demand growth pattern, again on the assumption that both lags are the same at any time, t, (see Fig. 3.1) f_t – the forecast or estimate of μ for the current period – can be estimated as:

$$f_t = u_t + (u_t - \bar{u}_t)$$

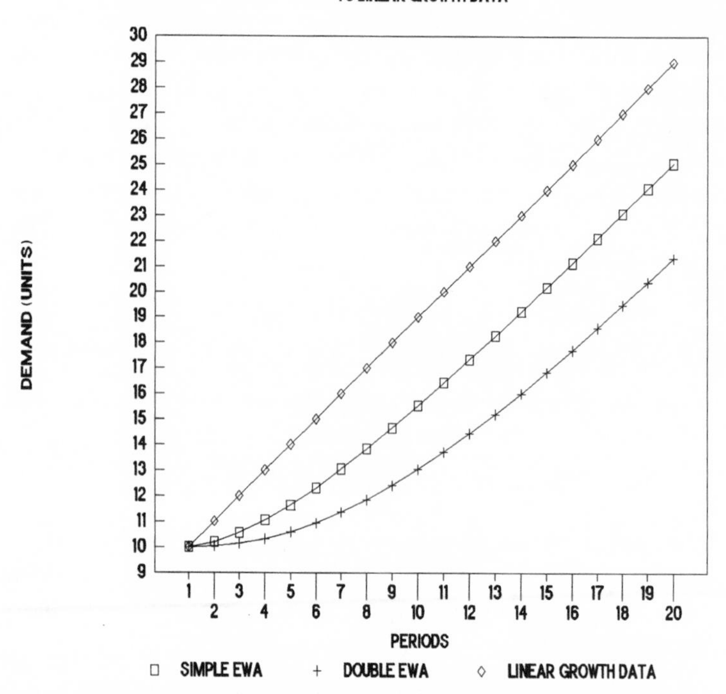

3.1 Response of a simple EWA, u_t, and a double smoothed EWA, $\bar{u}_t$, to linear growth demand data, d_t, with the simple EWA lagging the demand data by the same amount that the double smoothed EWA lags the simple EWA.

from which it follows that the forecast for T periods ahead, f_{t+T} is given by:

$$f_{t+T} = f_t + b_t T = 2u_t - \bar{u}_t + \frac{\alpha(u_t - \bar{u}_t)T}{(1-\alpha)} \qquad [3.3]$$

The improved performance of Brown's double smoothed model, as defined by Equation [3.3], compared with that of the simple exponentially weighted average model defined by Equation [3.2] can be seen in Fig. 3.2 which is generated by FOREMAN's TREND file using a longer version of the demonstration data POSIRAMP.

The main advantage of Brown's double smoothed growth model is that only one exponential weighting constant α is required. This simplifies the search procedure required to find the optimal value of α which minimises the Mean Squared Error.

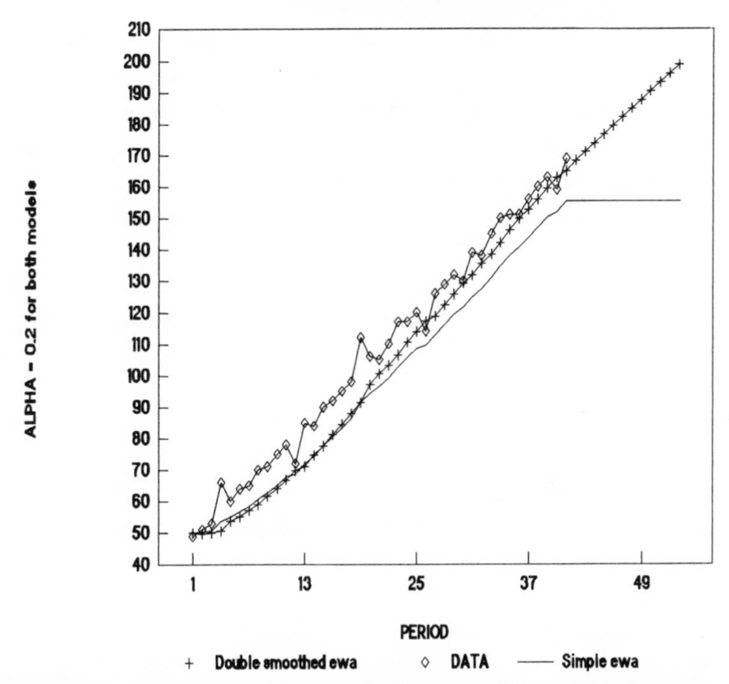

3.2 Response of Brown's double smoothed exponentially weighted average forecasting model with α fixed at 0.2 as generated by FOREMAN'S TREND file using the demonstration data POSIRAMP. The author has superimposed the exponentially weighted average forecasting model with α also fixed at 0.2 for comparison.

Task 3.1 Using FOREMAN'S TREND file establish for the DEMONSTRATION data series POSIRAMP the best value of α for Brown's double smoothed forecast. Justify your choice.

Holt's two parameter model

Holt's[2] proposal for an exponentially weighted average forecasting model capable of coping with growth data also accepts the fact that a simple EWA lags behind a growth trend but proposes that b_t – the estimate of the growth factor λ – can be evaluated as the exponentially weighted average of the difference between the current and immediate past simple exponentially weighted averages, hence:

$$b_t = \beta(u_t - u_{t-1}) + (1 - \beta)b_{t-1} \qquad [3.4]$$

where β is an exponential weighting constant which must take a value between 0 and 1.

In the Holt approach, the lag effect of the simple exponentially weighted average is overcome by a correction factor incorporated into the formula for calculating u_t – the estimate of the stationary element μ. This correction factor is implemented by adding the immediate past value of the estimate of the growth factor, i.e. b_{t-1} to the previous estimate of the stationary element of the forecast, which in this case is u_{t-1}. Hence u_t is now defined as:

$$u_t = \alpha d_t + (1 - \alpha)(u_{t-1} + b_{t-1}) \quad [3.5]$$

The forecast for T periods ahead, f_{t+T}, is then defined as:

$$f_{t+T} = u_t + b_t T \quad [3.6]$$

Although the Holt growth model based on Equations [3.4], [3.5] and [3.6] can operate as successfully as Brown's double smoothed model, because a two dimensional search procedure involving both the parameters α and β has to be undertaken to find the best forecast based on minimising the Mean Squared Error, Brown's single parameter model is generally preferred.

The Holt growth model is included here mainly because it is used to underpin the Holt Winters' model designed to cope with both growth and seasonality.

Seasonal forecasting

Where a strong seasonal influence is expected, one approach to forecasting is to assume that the seasonal pattern is defined by a set of 'de-seasonalising factors' with one for each period (usually month, quarter or planning period) within the overall seasonal cycle (usually a year). Thus for the twelve calendar months in a year, twelve de-seasonalising factors are required and these could be defined as the ratio between the expected demand for each month and the estimate of the stationary element. As an example, the set of de-seasonalising factors in Table 3.1 could be representative of a product with a strong demand during the summer and low demand during the winter.

Note that for a monthly situation the sum of the twelve ($L = 12$) de-seasonalising factors must sum to twelve – a necessary pre-requisite if unbiased forecasts are to be produced.

Table 3.1 Set of L = 12 de-seasonalising factors characterising data with high demand during summer months and low demand during winter months

Jan 1995 (t–L)	Feb	Mar	Apr	May	Jun	Jul	Aug	Sep	Oct	Nov	Dec	Jan 1996 (t)
0.7	0.6	0.6	0.9	1.1	1.3	1.4	1.4	1.1	1.0	1.0	0.9	
F_{t-L}	F_{t-11}	F_{t-10}	F_{t-9}	F_{t-8}	F_{t-7}	F_{t-6}	F_{t-5}	F_{t-4}	F_{t-3}	F_{t-2}	F_{t-1}	F_t

The major task in seasonal forecasting is to establish the values of the assumed de-seasonalising factors from a situation where, if no initial information as to the shape of the seasonal pattern is available, all values are assumed to have a value of one. Seasonal forecasts are then evaluated by multiplying non-seasonal forecasts by the appropriate de-seasonalising factor.

Holt Winters' model

The Holt Winters'[12] exponentially weighted average forecasting model for growth and seasonality assumes that demand values are drawn from a population composed of:

- a stationary element which exhibits no growth or seasonality but which might include step changes or single period impulses;
- a growth element in addition to the stationary element; and
- a seasonal element in addition to a growth and stationary element.

Stationary element

The identification of μ, the parameter describing the stationary element of Holt Winters' model, is essentially the same as that defined by Equation [3.5], with a modification due to the assumed seasonality of the current demand value, d_t. For instance by referring to Table 3.1 and assuming current time t is January, the value of the most recent de-seasonalising factor for January would be that which was established in January 1995 i.e. $F_{t-L} = 0.7$. This would infer that the current demand value, d_t, could be expected to be only 70% of that which would be expected in a typical, non-seasonal month. On the basis that the stationary element of the de-composition process is indeed attempting to describe the non-seasonal average demand value, the latest demand value d_t can be 'de-seasonalised' by dividing by $F_{t-L} = 0.7$ hence the estimated average of the stationary element u_t is given by:

$$u_t = \alpha\left(\frac{d_t}{F_{t-L}}\right) + (1-\alpha)(u_{t-1} + b_{t-1}) \qquad [3.7]$$

Growth element

The growth element in the Holt Winters' forecasting model is estimated as the exponentially weighted average of the difference between the current value of u_t and the immediate past value u_{t-1} as defined by Equation [3.4].

Seasonal element

The seasonal element of the Holt Winters' model is established as a set of L, de-seasonalising factors F_t, F_{t-1} ... F_{t-L}. The current value of the latest de-seasonalising factor F_t is evaluated as the exponentially weighted average of the ratio of the current demand value, d_t, to the most recently estimated value of the stationary element, u_t, hence:

$$F_t = \gamma\left(\frac{d_t}{u_t}\right) + (1-\gamma)F_{t-L} \qquad [3.8]$$

Within the three equations describing the Holt Winters' model, the exponential weighting constants, α, β and γ, clearly must all adopt values between 0 and 1. To obtain an optimal forecast it is necessary to undertake a complicated three dimensional search procedure, but a combination of values of 0.2, 0.05 and 0.5 respectively has been found to produce reasonably consistent good forecasts.

The response of the Holt Winters' forecasting model to a set of seasonal data is shown in Fig. 3.3 as generated by FOREMAN's TREND file and demonstrates the improvement in the forecast response as more years of data are processed. Because of the relative slowness of the exponential weighted average process in identifying the seasonal pattern through the de-seasonalising factors, it is usually recommended that at least four or five seasons of data be available for this model.

Task 3.2 From any source establish (using the DATAHELP file if necessary) a data file of a series of at least 36 monthly observations subjected to a seasonal influence. Using the forecasting models offered in the FOREMAN file TREND analyse the data, comment on the results and indicate if any of the models produce sensible forecasts. Justify your choice of best forecasting model.

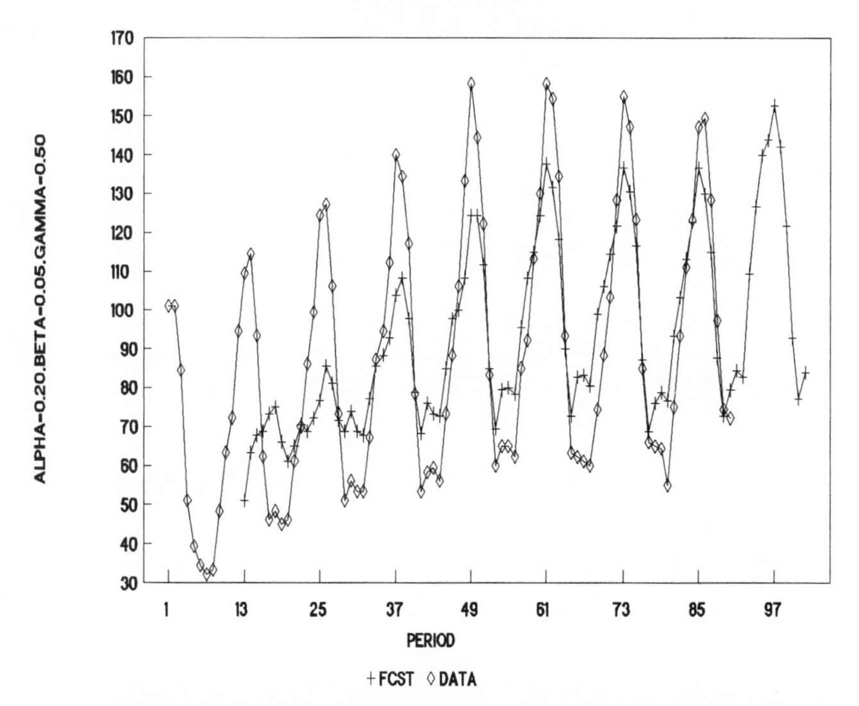

3.3 Response of Holt Winters' exponentially weighted average forecasting model with $\alpha = 0.2$, $\beta = 0.05$ and $\gamma = 0.5$ as generated by FOREMAN's TREND file using the demonstration data SEASONAL.

Conclusion

When the demand data exhibit growth or seasonal characteristics, the forecasting models required to achieve good results become more complex than the simple exponentially weighted average. This chapter has described the Brown's double smoothed and Holt's two parameter models, both of which are based on the exponentially weighted average concept but, unlike the simple exponentially weighted average, are designed to cope with growth. Similarly the Holt Winters' model is designed to cope with both growth and seasonality.

File from OPSCON's package FOREMAN associated with this chapter

TREND file – a multi-level menu driven file which allows the user to load either a file, or named range within a file of any suitable demand data or

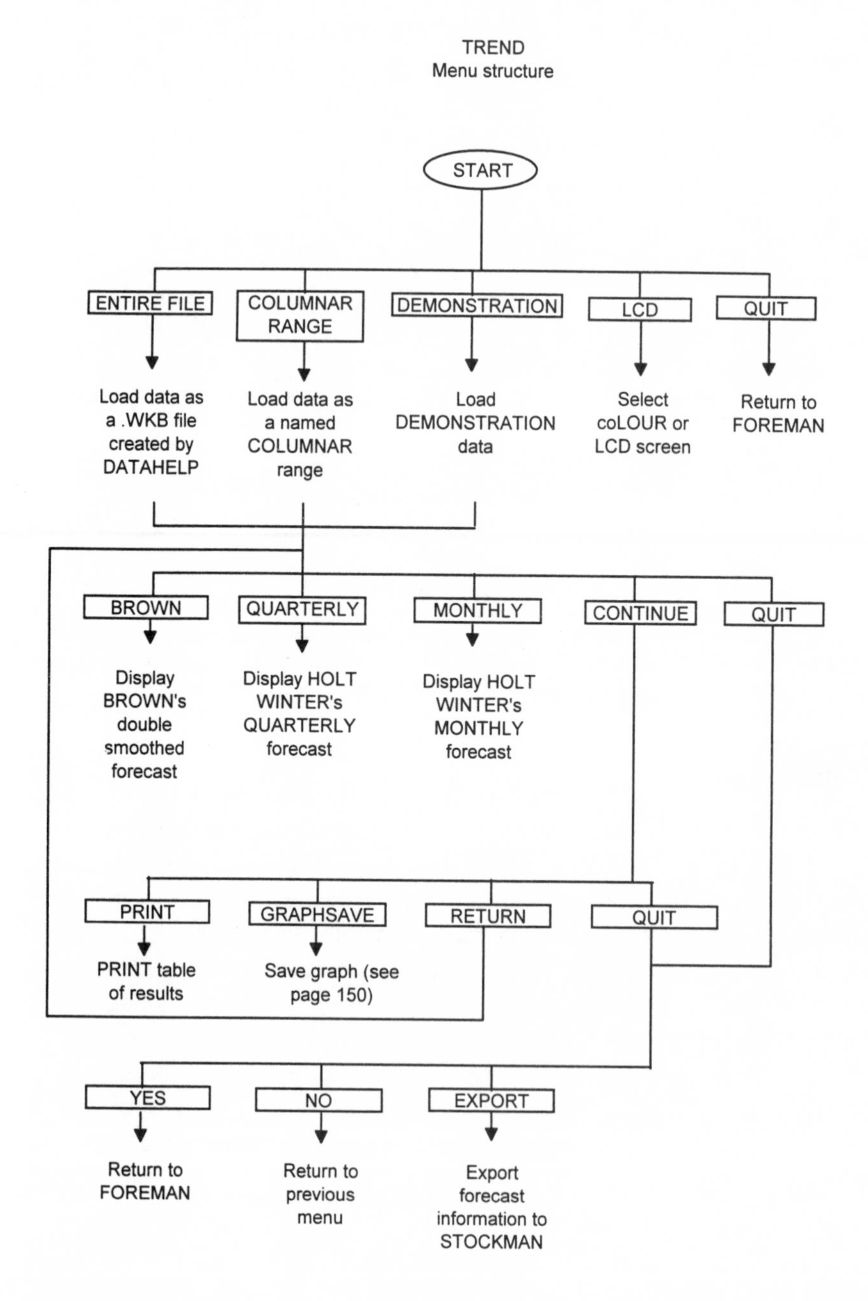

three demonstration sets of demand data. The following forecasts are established after an initial forecast value is supplied, namely:

- Brown's double smoothed exponentially weighted average forecasting model for growth;
- Holt Winters' exponentially weighted average forecasting model for growth and seasonality based on quarterly data; and
- Holt Winters' exponentially weighted average forecasting model for growth and seasonality based on monthly data.

For all models there is an option for specifying the values of α, β and γ.

The MSE and MAPE statistics are provided and options are available to print tables of results or save graphical PIC files of the forecasts' responses.

4

Medium-term forecasting for growth and seasonality

Introduction

This chapter describes the use of regression techniques to develop a group of curve fitting forecasting models. Although such medium-term, time series forecasting models are generally more sophisticated and more demanding in terms of computer processing time and storage than forecasts based on the exponentially weighted average concept, they do produce more information regarding the success or otherwise of the fitting process and also of the underlying composition of the demand data being analysed.

Forecasting methods which fall within the curve fitting definition attempt to identify a mathematical trend (or curve) which describes the underlying trend of the demand data being considered. Having identified best parameters (or coefficients) of the equation defining the trend through a fitting process, that equation can then be used to extrapolate into the future, thus producing forecasts beyond the end of the data. Where seasonal influences are present, the identification of the de-seasonalising factors describing that seasonality are established subsequent to identifying the trend.

Because of the more complicated calculation process involved with curve fitting forecasting models, together with the fact that all previous values used in the fitting process have to be stored, these models have not traditionally been used in conjunction with inventory control. However, more recently with increasing computing power and storage facilities at decreasing costs, such curve fitting forecasting models are now being associated with the more sophisticated inventory control packages together with forecasting models developed using such techniques as Fourier analysis and Baysian methods.

Regression and time series analysis

Regression in its broadest term is a technique which, when fitting a mathematically defined trend to a time series of observations, can identify directly those parameters (or coefficients) of the equation defining that trend which minimises the Mean Squared Error on the basis that all values within the series are of equal weight or importance. In the simplest case of a straight line model, the regression approach identifies the line of best fit which can then be extrapolated beyond the known data to produce forecasts.

Straight line trend forecasts

The simplest curve which could be considered for fitting as a trend forecasting model to a series of demand values is the linear curve (more familiarly the straight line) of the general form:

$$f_t = a + bt \qquad [4.1]$$

where:

- f_t is the estimate of the demand value at time period, t;
- the parameter **a** is calculated using simple regression analysis as applied to the n demand values making up the series. More specifically, **a** is the estimate of d_t at time zero; and
- the parameter **b** is the estimate of the slope of the trend line – also derived from regression analysis.

Having established the values of both the parameters **a** and **b** using standard linear regression techniques, the straight line defined by Equation [4.1] can be extended beyond the known values to provide forecasts of the presumed future demand.

A straight line forecasting model as defined by Equation [4.1], or line of best fit, would appear to be a good candidate for a medium-term trend forecasting model in business situations if it could be assumed that future demand values were likely to be linearly related to the passing of time (i.e. increasing by a fixed amount per period).

An example of fitting a straight line forecasting model to the DEMONSTRATION demand time series LINEAR using FOREMAN's STREXPO file is shown as Fig. 4.1.

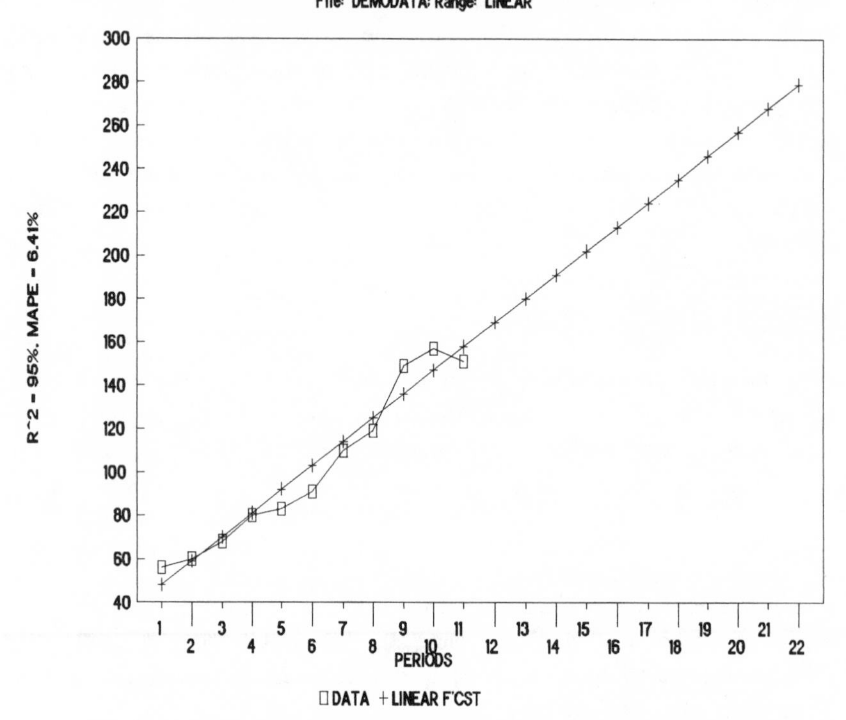

4.1 A straight line trend forecast fitted by FOREMAN's STREXPO file using the DEMONSTRATION data LINEAR.

Exponential growth trend forecasts

Although the linear or straight-line forecast described above is certainly a popular trend forecasting model, clearly if the observations under consideration do not exhibit a linear relationship with time, such a forecast will be inappropriate. Another relationship which tends to occur for time related data is that of exponential growth, where observed values tend to increase as a percentage of previous values, rather than by a fixed amount. Inflation operates in such a manner and, as such, influences many of the business variables such as customer demand that one might be attempting to forecast.

A simple exponential growth curve (not to be confused with a simple exponentially weighted average) which could be used for forecasting ahead could be defined as:

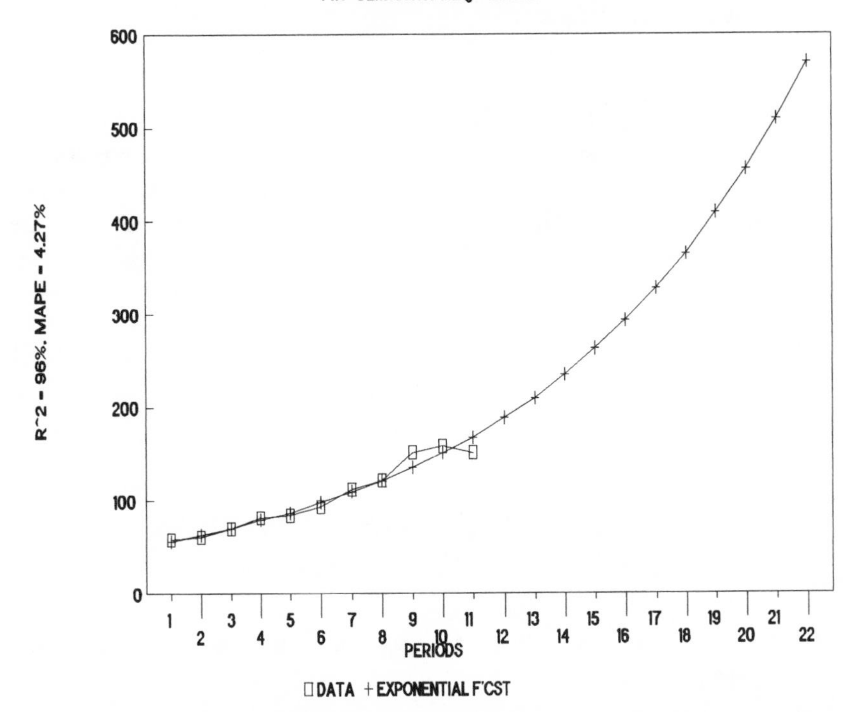

4.2 An exponential growth trend forecast fitted by FOREMAN's STREXPO file using the DEMONSTRATION data LINEAR.

$$f_t = ae^{bt} \quad [4.2]$$

where again **a** and **b** are both parameters whose values are established using regression but **e** is a special constant, namely the value of the base for natural logarithms:

$$e = 2.71828$$

An example of an exponential growth curve trend forecasting model fitted to the DEMONSTRATION demand time series LINEAR using FOREMAN's STREXPO file is shown as Fig. 4.2.

If the exponential growth curve can be considered to be linked closely to inflation, the parameter **b** would represent the equivalent of the rate of inflation. A value of **b** = 0.2 would then represent a 20% rate of inflation where the forecast of observed values would approximately double in three years and quadruple in seven. While such rates of growth (i.e. unit increases in time being accompanied by fixed, proportional increases in the dependent variable) cannot usually be sustained for long periods, over a forecast horizon of three to four years such a curve can produce reasonably accurate forecasts.

Forecasts based on polynomial equations

In many demand situations, it is necessary to develop a trend forecasting model which is able to represent turning points. Polynomial equations can be shown to produce one less turning point than the order of the polynomial. Hence a forecasting model based on a second order polynomial, as defined by Equation [4.3], produces a single turning point:

$$f_t = a + bt + ct^2 \qquad [4.3]$$

Where two turning points are thought more suitable, a third order polynomial as defined by Equation [4.4] produces two turning points.

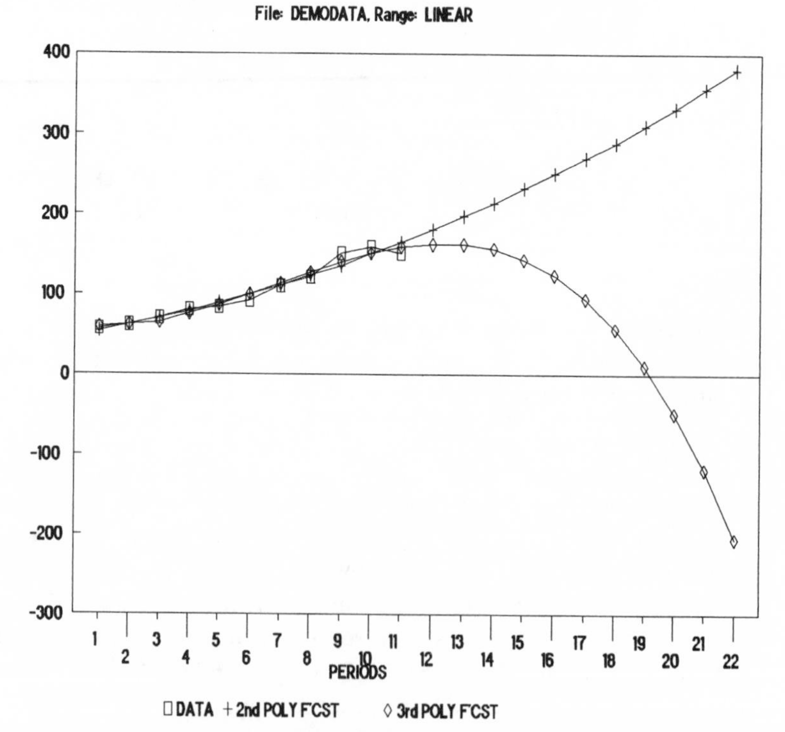

4.3 Second and third order polynomial trend forecasts fitted by FOREMAN's POLY2_3 file using the DEMONSTRATION data LINEAR.

$$f_t = a + bt + ct^2 + dt^3 \qquad [4.4]$$

An example of both second and third order polynomial curve trend forecasting models fitted to the DEMONSTRATION demand time series LINEAR using FOREMAN's POLY2_3 file is shown as Fig. 4.3.

Task 4.1 From any source establish (using the DATAHELP file if necessary) a data file of a series of at least 20 observations. Using the four trend curve forecasting models offered in the FOREMAN files STREXPO and POLY2_3 analyse the data, comment on the results and indicate if any of the models produce sensible forecasts. Justify your choice of best forecasting model.

Regression statistics for establishing the validity and accuracy of forecasting models

In addition to the Mean Squared Error (MSE) and Mean Absolute Percentage Error (MAPE) (see page 27 for more details), when using regression techniques it is possible to produce a further statistic known as the Coefficient of Determination which adds useful information regarding how well the selected forecasting model fits the demand data.

Coefficient of Determination

The Coefficient of Determination, R^2, developed from linear regression analysis, measures that proportion of total variation observed in the data which is explained by the trend curve being fitted. In the demand time series forecasting situation under consideration here, this definition can be extended to cover trend curves in general, such that:

$$R^2 = \frac{\text{Variability explained by forecast}}{\text{Total variability}}$$

If the trend curve forecast were to pass through all the observed demand values, then R^2 would be equal to 1 since all the variability in the data would then be explained by the curve. In practice this could never happen and, as a rule of thumb, most forecasters would require as a first condition an R^2 value of at least 0.85 (i.e. 85% of the variability in the data was explained by the fitted forecast) before any confidence would be assigned to the forecasts produced by the trend curve.

Where a degree of seasonal variation is known to exist, one would expect the proportion of variation explained by both the trend and seasonal models combined to be in excess of 85%.

When analysing a demand time series to establish whether forecasts beyond the end of the known data are likely to be reasonable predictions of the future, one would expect from the fitting procedure to have an R^2 value in excess of 0.85 and a MAPE value below, say, 20%. However, it should be remembered that both these statistics only indicate how well the proposed forecasting model fits the known data.

It should be noted, however, that when fitting polynomial curve forecasting models to a series of observations, the value of R^2 always increases as the degree of the polynomial increases. Hence a third order polynomial will always appear to produce a better fit than a second order polynomial and a second order polynomial will always appear to produce a better fit than a first order polynomial i.e. a straight line.

It should also be remembered that all forecasts into the future are based on the assumption that the characteristics displayed by the existing data will continue to influence future values. If this assumption does not apply, even with statistics proving a good fit to known data, forecasts ahead could be most inaccurate. In selecting an appropriate forecasting model it is therefore important to study the pattern of forecasts produced ahead of the known data and not to rely solely on the model which produces the highest value of R^2. This is exemplified by the third order polynomial forecast in Fig. 4.3 which although producing a high R^2 value produces unacceptable forecasts.

Time series analysis – forecasting: seasonality

One approach to forecasting seasonal demand is to assume initially that there could be an underlying medium-term trend and to identify this trend using traditional regression techniques as discussed earlier in this chapter. Having established whether or not such a trend exists, if subsequently the Coefficient of Determination, R^2, is still very low (i.e. less than 0.85) and it is therefore apparent that a considerable amount of the data's variability still remains unexplained, it is worth examining the residuals (forecasting errors) to see whether that unexplained variability could be due to seasonal influences.

This overall concept of assuming that a demand series could be made up of a combination of underlying factors is often referred to as a decomposition approach within which it is assumed that the time series data can be made up of three elements, namely:

- a medium-term trend;
- a seasonal influence based on a well recognised time cycle, such as a calendar year; and
- superimposed noise or random variation with an assumed mean of zero.

This decomposition principle argues that having established the properties of both trend and seasonal elements of the data, these can then be recombined to produce a forecast.

De-seasonalising factors

A de-seasonalising factor (for earlier discussion on de-seasonalising factors see page 48) in the context of a ratio-to-trend forecasting model, represents the ratio of the value of an observation for a specific time period within a seasonal cycle to the expected value at that time, this latter being assumed in this situation to be the value of the previously established medium-term trend for that period.

For each period within a seasonal cycle, a raw de-seasonalising factor is established as the average of the ratios of the observed demand values to the trend forecast values for the corresponding period. Thus, if the ratios for the months of January for the three years 1993, 1994 and 1995 were 1.3, 1.7 and 1.5 respectively; then the raw de-seasonalising factor for January would be established as the average of these three values, namely 1.5. Having established raw de-seasonalising factors for all periods within the seasonal cycle (typically either for the four quarters or twelve months of a calendar year), these are then normalised to ensure that no bias is introduced to the forecasting process. Hence, the twelve de-seasonalising factors for a monthly forecasting model would be constrained to sum to twelve (12) by simply multiplying each raw de-seasonalising factor by twelve and dividing by the sum of the raw values. Similarly, quarterly de-seasonalising factors would be constrained to sum to four (4).

A typical set of de-seasonalising factors for data for which it was assumed there was a monthly seasonal pattern could be of the form shown in Table 4.1 indicating the demand for a product or service which peaks in the winter months and falls during the summer.

Because only one observation per year contributes towards the evaluation of de-seasonalising factors, to establish an underlying seasonal pattern a minimum of three seasons of data is generally required (i.e. 36 monthly or 12 quarterly demand values).

Table 4.1 Set of twelve de-seasonalising factors characterising data with high demand during winter months and low demand during summer months

Jan 1995	Feb	Mar	Apr	May	Jun	Jul	Aug	Sep	Oct	Nov	Dec
1.5	1.3	1.0	0.9	0.8	0.8	0.7	0.8	0.9	1.0	1.1	1.2

An example of fitting a monthly ratio-to-trend seasonal forecasting model based on a third order polynomial trend is shown in Fig. 4.4. It can be seen that the two turning points of the third order polynomial fitted as a trend curve appear to represent well the underlying trend of the data even though it can be seen from the analysis shown in Fig. 4.4 that this trend only explains 14% (i.e. R^2 = 0.14) of the underlying variability in the data. Subsequent application of the ratio-to-trend seasonal forecasting model can be seen to bring this up to a total of 98% (i.e. R^2 = 0.98) if monthly data is assumed. From this it can be concluded that:

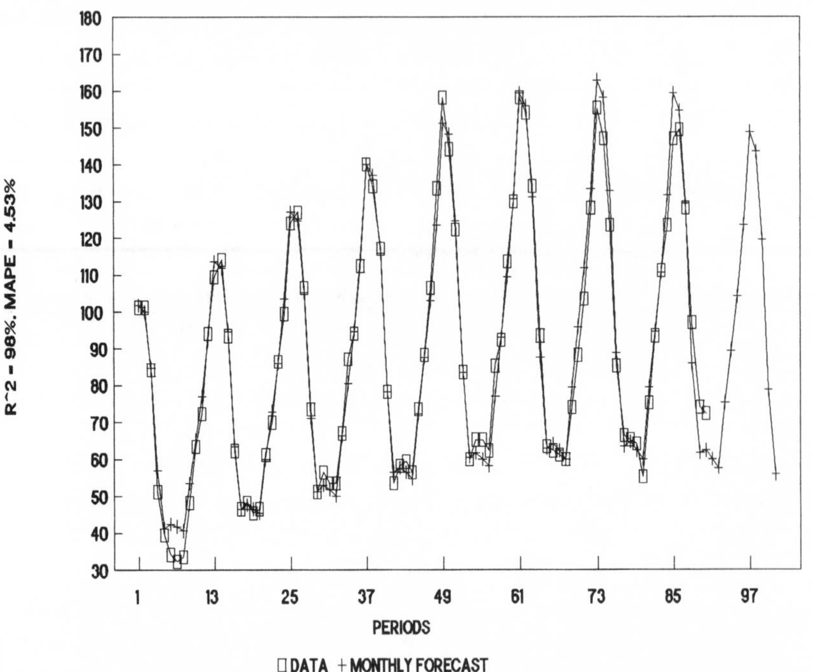

4.4 A third order polynomial trend curve together with a monthly ratio-to-trend seasonal forecasting model fitted to the DEMONSTRATION data SEASONAL using FOREMAN's POLY2_3 file.

Task 4.2 From any source establish (using the DATAHELP file if necessary) a data file of a series of at least 36 monthly observations subjected to a seasonal influence. Using the four trend curve forecasting models offered in the FOREMAN files STREXPO and POLY2_3 analyse the data, comment on the results and indicate if any of the models produce sensible forecasts. Justify your choice of best forecasting model.

- 14% of the data variability is explained by the third order polynomial trend curve forecasting model;
- 84% (i.e. 98% minus 14%) of the data variability is explained by a monthly ratio-to-trend seasonal forecasting model; and
- 2% of the data variability (i.e.100% minus 98%) remains unexplained and is presumed to be made up of errors distributed normally and with a mean of zero.

Considered together, Fig. 4.4 and 4.5 depict a very successful forecasting analysis of a relatively complex demand pattern.

C.A.L. Module OPERATIONS CONTROL – Forecasting CDL/96

3rd POLY QUARTERLY MONTHLY FORMULAS CONTINUE

Plot trend curve alone

File: DEMODATA, Range: SEASONAL 90 HORIZON: 12

PERIOD t	DATA y	3rd POLY Y	QLY SEASON	MONTHLY SEASON
1	101	63.14	65.69	101.49
2	101	63.70	69.19	99.66
3	84	64.26	64.22	84.35
4	51	64.84	56.66	56.74
5	39	65.43	68.08	41.08
6	34	66.03	71.73	42.17
7	32	66.64	66.60	41.23
8	33	67.26	58.70	40.00
9	48	67.89	70.64	53.41
10	63	68.53	74.44	64.75
11	72	69.17	69.13	76.57
12	94	69.83	61.02	92.24
13	109	70.48	73.34	113.29
14	114	71.15	77.29	111.32
15	93	71.82	71.78	94.27

Seasonal time series analysis

COEFFICIENT OF DETERMINATION R^2	
3rd POLY	0.14
QUARTERLY	0.13
MONTHLY	0.98

MEAN ABSOLUTE PERCENTAGE ERROR	
3rd POLY	35.64%
QUARTERLY	36.04%
MONTHLY	4.53%

File:POLY2_3 Esc toggles menu MENU

4.5 FOREMAN's POLY2_3 file's screen display when a third order polynomial trend curve together with a monthly ratio-to-trend seasonal forecasting model has been fitted to the DEMONSTRATION data SEASONAL using FOREMAN's POLY2_3 file.

Conclusion

Forecasting models based on curve fitting procedures, although more demanding in computational and storage terms than models based on the exponentially weighted average concept, can provide more information concerning the underlying character of the data being analysed.

This chapter has considered the use of the following trend curves, namely:

- a straight line, $f_t = a + bt$;
- an exponential growth curve, $f_t = ae^{bt}$;
- a second order polynomial, $f_t = a + bt + ct^2$; and
- a third order polynomial, $f_1 = a + bt + ct^2 + dt^3$.

For all these models, the value of the Coefficient of Determination R^2 indicates that proportion of the overall variability of the data which is explained by the fitting of the curve. Where the value of R^2 is considerably less than 0.85 the residuals (forecasting errors) should be examined for evidence of seasonality. Where such seasonality is shown to exist, a ratio-to-trend seasonal model may bring the value of R^2 for the combination of the trend and seasonal forecasting above the desired value of 0.85.

Ideally as well as a high value of R^2 , the Mean Absolute Percentage Error (MAPE) should be low (see Table 2.2, page 28 for recommended values).

Files from OPSCON's package FOREMAN associated with this chapter

STREXPO and POLY2_3 files

Two multi-level menu driven files with identical structures allow the user to load either a file, or named range within a file, of any suitable demand data or three demonstration sets of demand data. The following trend forecasts are available, namely:

STREXPO

- a straight line, $f_t = a + bt$;
- an exponential growth curve, $f_t = ae^{bt}$.

POLY2_3

- a second order polynomial, $f_t = a + bt + ct^2$;
- a third order polynomial, $f_t = a + bt + ct^2 + dt^3$.

Having initially fitted a trend curve, for all models there is a subsequent option which allows for the fitting of either a quarterly or monthly seasonal ratio-to-trend forecasting model.

The MSE and MAPE statistics are provided for both the trend curve forecast and the subsequent seasonal forecasts. Options are available to print tables of results or save graphical PIC files of the forecasts' response.

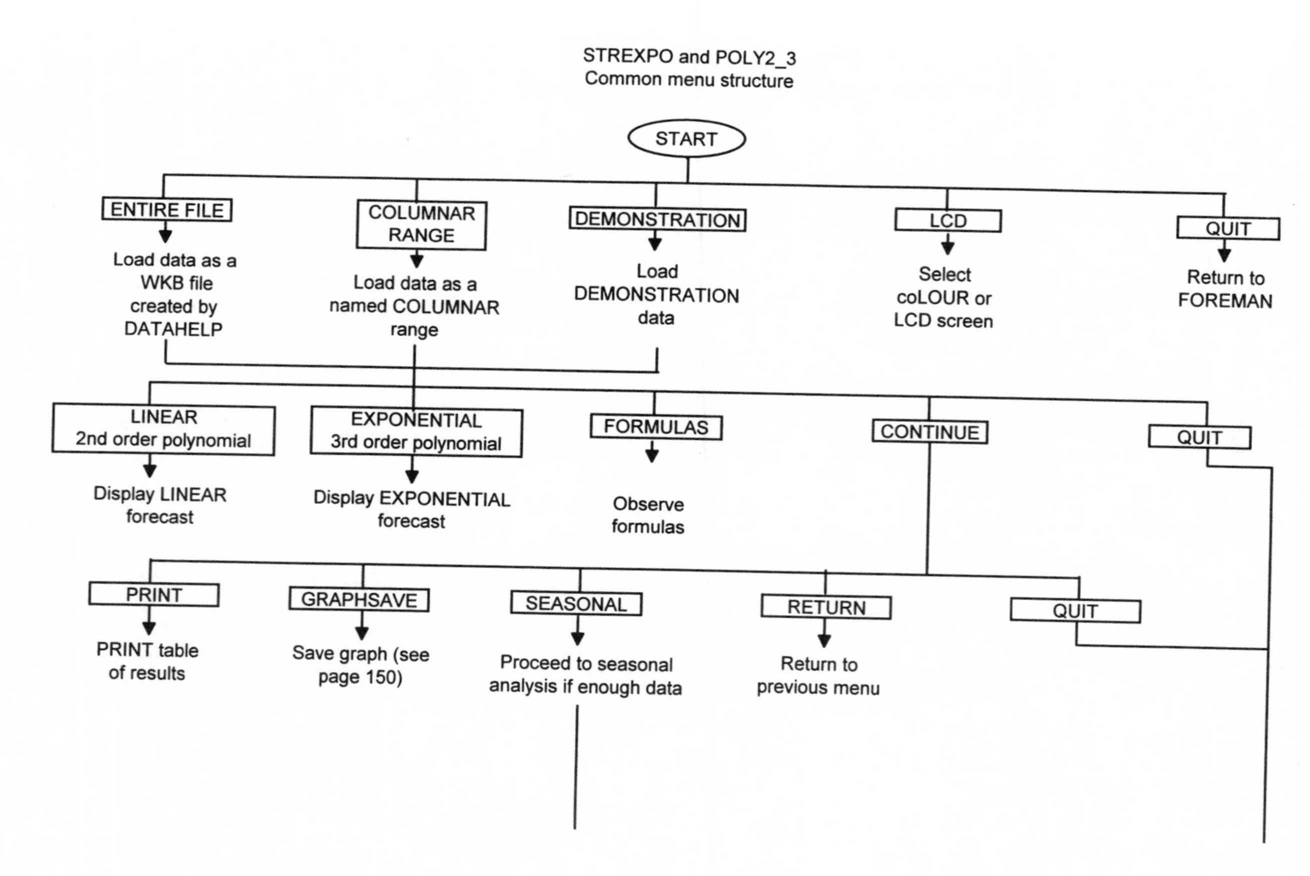
STREXPO and POLY2_3
Common menu structure
START
ENTIRE FILE
Load data as a WKB file created by DATAHELP
COLUMNAR RANGE
Load data as a named COLUMNAR range
DEMONSTRATION
Load DEMONSTRATION data
LCD
Select coLOUR or LCD screen
QUIT
Return to FOREMAN
LINEAR
2nd order polynomial
Display LINEAR forecast
EXPONENTIAL
3rd order polynomial
Display EXPONENTIAL forecast
FORMULAS
Observe formulas
CONTINUE
QUIT
PRINT
PRINT table of results
GRAPHSAVE
Save graph (see page 150)
SEASONAL
Proceed to seasonal analysis if enough data
RETURN
Return to previous menu
QUIT

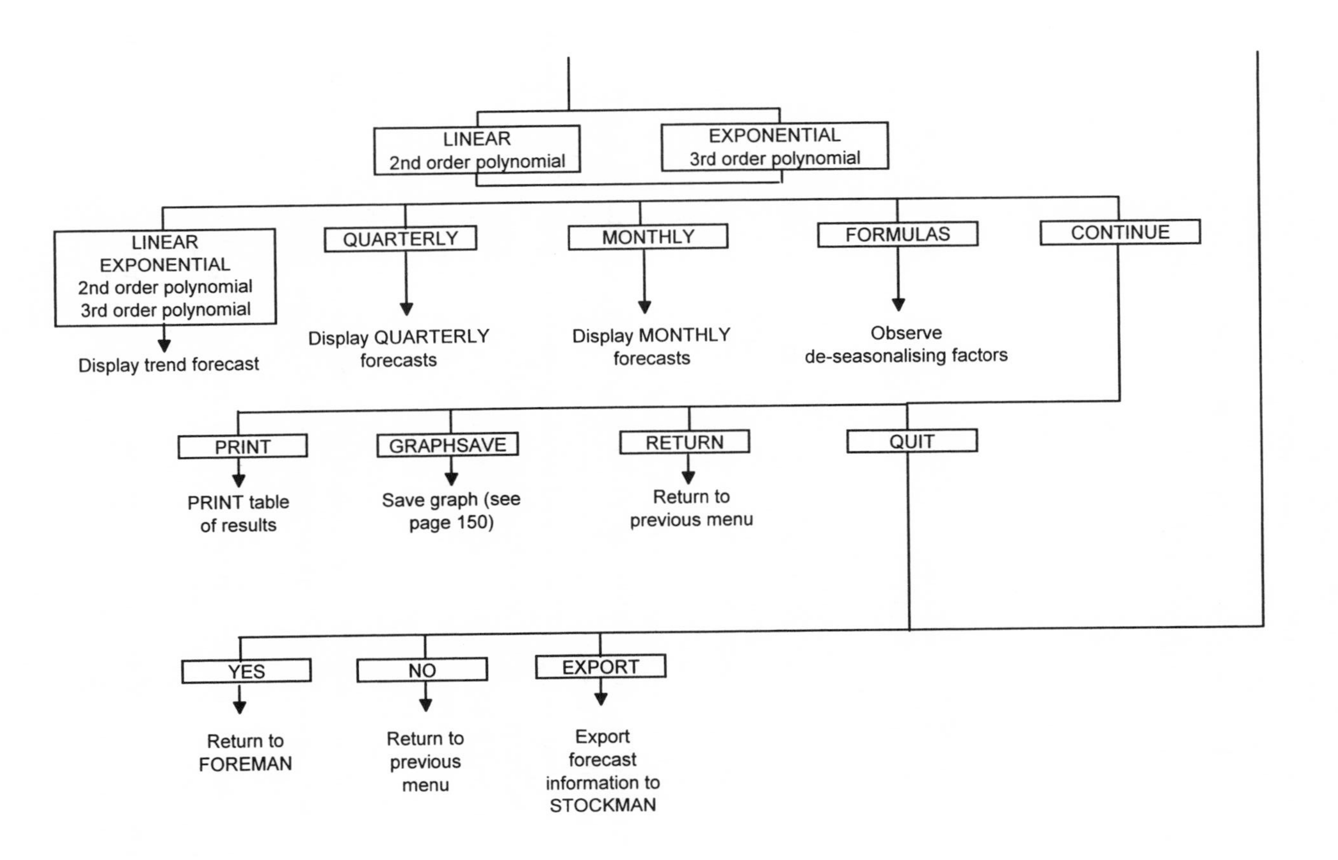
LINEAR
2nd order polynomial
EXPONENTIAL
3rd order polynomial
LINEAR
EXPONENTIAL
2nd order polynomial
3rd order polynomial
Display trend forecast
QUARTERLY
Display QUARTERLY forecasts
MONTHLY
Display MONTHLY forecasts
FORMULAS
Observe de-seasonalising factors
CONTINUE
PRINT
PRINT table of results
GRAPHSAVE
Save graph (see page 150)
RETURN
Return to previous menu
QUIT
YES
Return to FOREMAN
NO
Return to previous menu
EXPORT
Export forecast information to STOCKMAN

5

The essential links between forecasting and inventory control

Introduction

Since the fundamental purpose of holding inventory is to act as a buffer between demand and supply, it is essential when designing an inventory system that the parameters describing the demand situation be known. In practice, for most reasonably fast moving items, it can usually be assumed that the distribution of demand per unit time is normal, and that the demand situation is therefore fully defined by the average, $\bar{D}$, and the standard deviation σ_d. On the supply side, it is necessary to know the duration of the replenishment leadtime, L, i.e. the delay between the raising of a replenishment order and its subsequent receipt. For the re-order cycle inventory policy in particular, it is also necessary to know the review period, R, i.e. the interval at which stocks are reviewed such that a new order for replenishment can be raised.

In terms of selecting an appropriate forecasting method to associate with particular stocked items held within an inventory system, there are essentially three options on offer and hence the Pareto approach, which divides stocked items into types 'A', 'B' and 'C', is considered an appropriate categorisation system for recommending which type of forecast model is most appropriate for each type of stocked item.

The essential links between forecasting and inventory control

The majority of inventory control models for relatively fast moving items (i.e. those with an average demand per unit time of the order of 20 units or more) assume that the demand per unit time is distributed normally and that the distribution is therefore completely defined by:

- $\bar{D}$ – the average demand per unit of time (usually equated with the appropriate forecast i.e. $\bar{D} = f_{t+T}$). The higher the average demand, the more stock that will be required to meet the average demand during the leadtime.

- σ_d – the standard deviation of demand per unit of time which for short-term forecasting models would usually be equated as the Mean Absolute Deviation of forecasting errors multiplied by 1.25 but which for medium-term forecasting models would normally be evaluated as the square-root of the Mean Squared Error. A higher variability of demand (and hence a larger value of the standard deviation of demand) requires more stock to absorb those variations if a reasonable level of service is to be achieved.

In practice, for most computerised stock control systems, the moment a new forecast is generated both the average demand, $\bar{D}$, and standard deviation of demand, σ_d, are updated in line with the new forecast. This also means that the inventory control system parameters, such as the re-order level and replenishment quantity (for the re-order level policy) and maximum stock level (for the re-order cycle policy) are also updated in line with the fresh forecast information. This updating feature of inventory control parameters in line with changes in forecasting information is simulated within the OPSCON package. Thus, at the conclusion of any of the forecasting analyses within FOREMAN an option is available to save the result of a particular analysis temporarily before proceeding directly to STOCKMAN's inventory policy simulation files. Here, when specifying the demand characteristics, the FORECASTING option automatically uses these results to specify the average and standard deviation of demand.

Selection of forecasting models based on type of stocked item – Pareto analysis

Within any inventory control system involving many stocked items it is clearly not sensible to treat all items as equally important in terms of the selection of an appropriate model type to forecast demand, of which, broadly speaking, there are three different approaches, namely:

1. Non-adaptive forecasting models. For the short-term, exponentially weighted family of forecasting models this includes those forecasting models within which the exponential smoothing constant(s) remain fixed whereas all the medium-term, curve fitting forecasting models fit into this category.
2. Adaptive forecasting models. For the short-term, exponentially weighted average family of forecasting models this includes those

forecasting models within which the exponential smoothing constant is allowed to vary in line with the characteristic of the data.
3. No forecast. An option which could be valid if the advantages gained from forecasting do not offset the costs involved in forecasting.

With these three broadly based options, it would be convenient to adopt a categorisation approach which also divided stocked items into three categories.

The most commonly used categorisation system for stocked items is based on the naturally occurring relationship which tends to be prevalent in many organisational situations where 'a minority of items represent a major proportion of measured importance and conversely a majority of items represent a minor proportion of measured importance'. This relationship, originally recorded by Pareto with regard to the small number of people within the population who owned a major proportion of the nation's wealth in eighteenth century Italy, has also been recognised within most inventory control situations ever since computers have been capable of sorting (or indexing) large numbers of stocked parts in descending order of value, in addition to the more orthodox sorting by alphabetic order of part number.

This Pareto relationship for a list of 250 stocked items is demonstrated in STOCKMAN's PARETO file within which stocked items can be presented either alphabetically or in descending order of value as stocked. When the PARETO file is initially loaded from the STOCKMAN menu, the 250 parts held in the list are ordered alphabetically by part number as can be seen in Fig. 5.1, which is a view of the PARETO file's initial screen where, in addition to part number:

- the quantity held in stock;
- the average unit cost;
- the value of stock at average cost; and
- the cumulative value of stock held (at average cost);

are also displayed.

Were all 250 items held in this stock list of equal value, clearly a plot of cumulative value would approximate to a straight, diagonal line. Given, however, that within the list there are a few high value items and many low value items, one would expect the cumulative value plot to approximate to a straight, diagonal line with a few sudden step changes caused by the occasional, valuable item. This can be confirmed by examination of Fig. 5.2 which is shown in response to the PARETO file's VIEWGRAPH option when stocked parts are sorted alphabetically by part number.

C.A.L. Module OPERATIONS CONTROL - Stock Control CDL/96
NUMBER STOCKVALUE VIEWGRAPH coLOUR PRINT QUIT
Sort alphabetically by part number

* PARETO DEMONSTRATION * [LISTED IN ALPHABETIC ORDER OF PART-NUMBER]

#	PART-NUMBER	QUANTITY HELD IN STOCK	AVERAGE UNIT COST	VALUE OF STOCK AT AVERAGE COST	CUMULATIVE VALUE OF STOCK HELD
1	ABC1432	700	34.20	23,940.00	23,940.00
2	ABC1433	700	34.00	23,800.00	47,740.00
3	ACR8907	70000	0.34	23,800.00	71,540.00
4	ACR8914	70000	0.67	46,900.00	118,440.00
5	ADE454	8638	4.00	34,552.00	152,992.00
6	ADE565	5355	5.00	26,775.00	179,767.00
7	ADE676	10500	2.00	21,000.00	200,767.00
8	ADG765	1306	58.00	75,759.60	276,526.60
9	ADR4545	10500	2.00	21,000.00	297,526.60
10	ADR876	19593	3.00	58,779.00	356,305.60
11	AFB653	1400	34.00	47,600.00	403,905.60
12	AGR876	19593	4.00	78,372.00	482,277.60
13	AGT3545	1306	54.00	70,534.80	552,812.40

File:PARETO Esc for menu MENU

5.1 The start of STOCKMAN's PARETO file's list of 250 stocked items listed in alphabetical order of part number.

Within the PARETO file, the STOCKVALUE option automatically sorts parts in descending order of stock value (at average cost) as shown in Fig. 5.3 to reveal (again through the VIEWGRAPH option) the typical Pareto or ABC relationship shown as Fig. 5.4 within which the:

- 'A' items comprise 10–20% of the items listed but represent of the order of 80% of the total value (hence the term '80/20 rule');
- 'B' items comprise 20–30% of the items listed and represent about 15–20% of the total value; and
- 'C' items comprise 50–70% of the items listed but only represent about 5–10% of the total value.

The shape of the Pareto curve, as seen in Fig. 5.4, will vary from organisation to organisation to the extent that some theorists would even claim in the case of manufacturing companies to be able to identify the type of product and product mix being made through the shape of the company's Pareto curve. However, in summary, Pareto or ABC analysis can be used to identify three classes, groups or categories of stocked items within virtually all inventory control systems. While accepting that ultimately all stocked items are 'important' irrespective of value, this approach identifies:

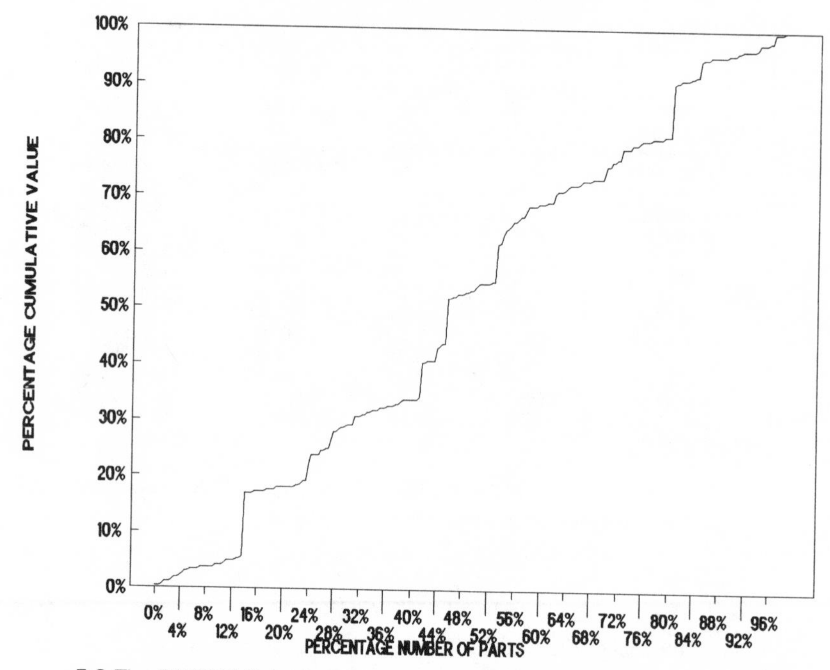

5.2 The PARETO file's plot of cumulative value of stock held (at average cost) when parts are listed in alphabetic order of part number.

- 'A' items which are relatively few in number and represent the major portion of capital investment in stocked items;
- 'B' items which occur in greater number than 'A' items but are of moderate value in terms of capital investment; and
- 'C' items which are the majority of items but only represent a very small proportion of capital investment in stocked items.

Choosing the most appropriate group of forecasting model(s) for stocked items

In this section, consideration will be given to which of the groups of forecasting models available within FOREMAN should be associated with which type of stocked items based on the ABC categorisation system.

C.A.L. Module OPERATIONS CONTROL - Stock Control CDL/96
NUMBER STOCKVALUE VIEWGRAPH colOUR GRAPHSAVE PRINT QUIT
Sort in descending order of value of stock

* PARETO DEMONSTRATION * [LISTED IN DESCENDING ORDER OF VALUE]

#	PART-NUMBER	QUANTITY HELD IN STOCK	AVERAGE UNIT COST	VALUE OF STOCK AT AVERAGE COST	CUMULATIVE VALUE OF STOCK HELD
1	DDH098	26124	80.00	2,089,920.00	2,089,920.00
2	JXC0987	3919	450.00	1,763,370.00	3,853,290.00
3	FYD098	17416	90.00	1,567,440.00	5,420,730.00
4	GHY67676	11756	110.00	1,293,138.00	6,713,868.00
5	FRT789	17634	67.00	1,181,457.90	7,895,325.90
6	DOL9432	827	600.00	496,356.00	8,391,681.90
7	KYU8989	784	600.00	470,232.00	8,861,913.90
8	FTR7878	823	543.00	446,837.96	9,308,751.86
9	HYU78787	3962	100.00	396,214.00	9,704,965.86
10	JFG0987	3919	100.00	391,860.00	10,096,825.86
11	DQA5454	8708	45.00	391,860.00	10,488,685.86
12	GHY787	4136	78.00	322,631.40	10,811,317.26
13	EFT6543	13062	23.00	300,426.00	11,111,743.26

File:PARETO Esc for menu MENU

5.3 The start of STOCKMAN's PARETO file's list of 250 stocked items listed in descending order of stock value at average cost.

Allocation of forecasting model(s) to 'A' items

Because 'A' items are few in number but of great value they should be associated with the highest degree of control. This suggests that the forecasts of demand for such items should be carefully monitored, which presupposes that such items should be monitored using the smoothed error tracking signal (see page 33). If these 'A' items' forecasts are to be monitored, this negates the use of the tracking signal as the exponential smoothing constant for short-term forecasting models. Hence it is proposed that, for 'A' items, non-adaptive forecasting models are most appropriate. Subsequent to this decision, which particular non-adaptive forecasting model is chosen must be a function of the type of demand characteristics each item exhibits as follows:

- stationary demand – simple exponentially weighted average forecasting model (see page 24);
- demand subject to growth – Brown's double smoothed exponentially weighted average forecasting model (see page 45) or any of the medium-term, curve fitting forecasting models as described in Chapter 4; and
- demand subject to growth and seasonality– Holt Winters' exponentially weighted average forecasting model (see page 49) or any of the medium-

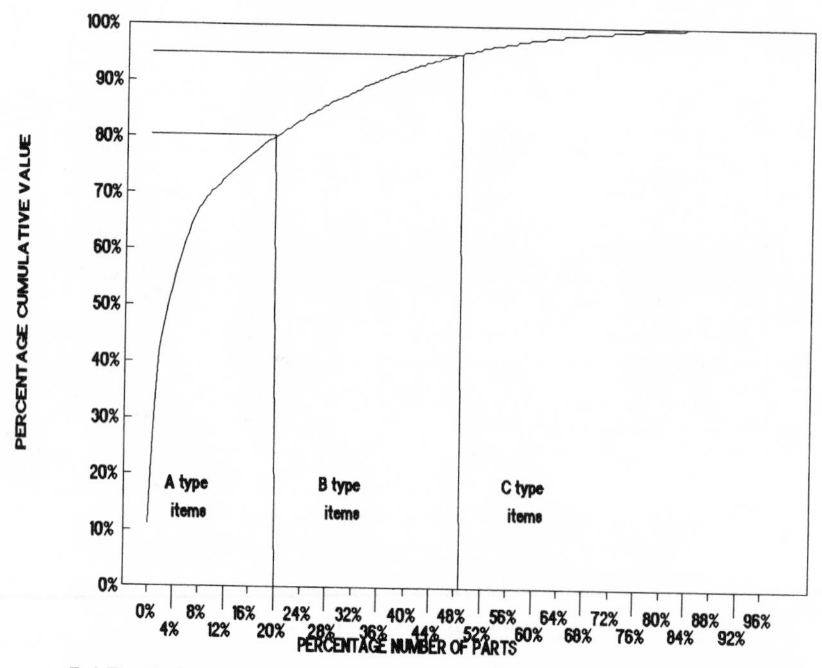

5.4 The PARETO file's plot of cumulative value of stock held (at average cost) when parts are listed in descending order of stock value.

term, curve fitting models with subsequent ratio-to-trend forecasting as described in Chapter 4.

All the forecasts should be monitored using the smoothed error tracking signal to detect when forecasts go 'out of control' to sudden changes in the demand pattern.

Allocation of forecasting model(s) to 'B' items

Because 'B' items are relatively numerous and of moderate value, on balance they should be associated with a group of forecasting models which can guarantee that changes in demand pattern are automatically accommodated by a responsive forecast rather than that such changes be coped with by manual intervention via a monitoring system. This suggests that such items could be successfully forecast using short-term, adaptive forecasting models. Of those adaptive forecasting models available, the delayed adaptive response model has gained greatest popularity with its ability to respond rapidly to genuine step changes in demand while ignoring single period impulses (see page 38). In cost benefit terms, the secondary importance of 'B' type items would appear to rule out all the more expensive medium-term, curve fitting forecasting models.

Allocation of forecasting model(s) to 'C' items

In considering the most appropriate group of forecasting models which could be associated with 'C' items, the fact that 'C' items represent the majority of items in stock control systems, but a very small proportion of overall value, must raise the issue of whether demand forecasts should be generated at all for such items. This suggestion may be anathema to those companies and organisations that produce forecasts for all stocked items irrespective of value but should be considered in the following context.

If 'C' items represent only 5% of the total value of stocked items, a possible solution to offering a reasonable level of service and avoiding stockouts could be simply to overstock by, say, 20%. Such an apparently preposterous proposal looks more sensible when related to the fact that a 20% increase on a 5% investment involves a cost of only 1%, and this cost should be set against the actual costs involved in operating even a relatively cheap, short-term forecasting procedure linked to a stock control system which, just from a forecasting point-of-view, would include the costs of:

- collecting the necessary demand data;
- calculating the appropriate forecast; and
- storing relevant information which for an exponentially weighted average forecasting model would be the forecast itself, the previous smoothed error and Mean Absolute Deviation (see pages 33 and 34) together with the associated value of the exponential smoothing constant.

Clearly there are significant costs involved in operating such a set of forecasting procedures and, on a 'cost benefit' approach, if these costs together with the costs of implementing the associated inventory control system exceed the approximate 1% increased cost involved in achieving control by a policy of overstocking by 20%, an operating policy of forecasting for all 'C' items irrespective of value is surely brought into question.

Conclusion

Forecasting demand is a necessary precursor to operating a successful inventory control system. The forecasting procedures must not only measure the likely future average demand but also the variation or spread of demand around that average value, as measured by the standard deviation. A method of allocation type of forecasting model to type of stocked item is to use Pareto analysis and to allocate on the basis of whether the item involved is categorised as an 'A', 'B' or 'C' item.

File from OPSCON's package STOCKMAN associated with this chapter

PARETO file – a simple, single level menu driven file which displays the characteristic relationship of multi-item stock control systems where the parts are listed either:

- in alphabetical order of part number; or
- in descending order of value.

Where stocked items are listed in descending order of value it is clear that they can be subdivided into three distinct categories as:

- 'A' items which are relatively few in number and represent the major portion of capital investment in stocked items;
- 'B' items which occur in greater number than 'A' items but are of moderate value in terms of capital investment; and
- 'C' items which are the majority of items but only represent a very small proportion of capital investment in stocked items.

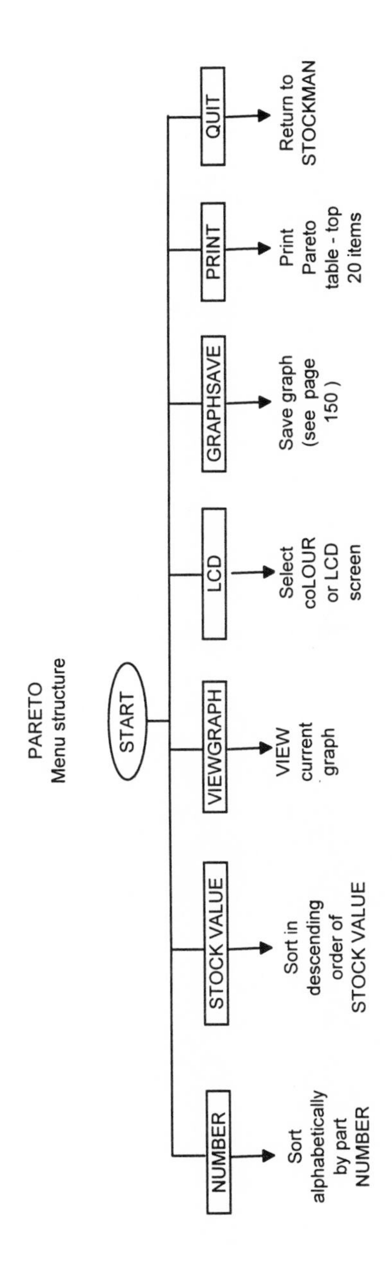
PARETO
Menu structure
START
NUMBER
Sort alphabetically by part NUMBER
STOCK VALUE
Sort in descending order of STOCK VALUE
VIEWGRAPH
VIEW current graph
LCD
Select coLOUR or LCD screen
GRAPHSAVE
Save graph (see page 150)
PRINT
Print Pareto table - top 20 items
QUIT
Return to STOCKMAN

Part II

Inventory control – the re-order level inventory policy

6

Establishing the value of the re-order level

Introduction

As was indicated originally in Chapter 1, to develop a successful inventory control system it is necessary to specify a set of rules to decide how such a system is to be replenished in the light of:

- the demand pattern that the inventory control system is subjected to; and
- the procurement leadtime – the delay between placing an order for replenishment on a supplier and its subsequent receipt into stock ready for issue.

The set of rules controlling the timing and size of the replenishment order then becomes an inventory policy. The two most commonly implemented inventory policies – from which virtually all others are derived – are:

- the re-order level policy (sometimes known as the fixed order quantity system); and
- the re-order cycle policy (sometimes known as the periodic review system).

This chapter is concerned only with the former, namely the re-order level policy, and within that policy evaluation specifically of the re-order level. Chapter 7 considers the evaluation of the replenishment order size within the re-order level policy and Chapter 8 investigates the interaction of both these policy parameters.

The re-order level (ROL) policy

The most commonly used inventory control policy is the re-order level policy, within which a replenishment order is placed when the stock-on-hand (i.e. physical stock held plus outstanding replenishments less committed stock) falls to or below a level known as the re-order level. When a replenishment order is raised, as a result of either equalling or falling below the re-order level, the size of the replenishment order placed is usually for a fixed amount. This replenishment order is delayed by a period known as the 'leadtime', this being the delay between the order being placed on a supplier and the goods being received by the organisation placing the order. To this must be added the additional delay which inevitably occurs within that organisation before the replenishment order is available to meet customers' demands via the inventory control system. For this policy the leadtime is the period of risk during which time stocks are at risk of running out since, within the rules of the policy, no remedial action can be taken to prevent this happening should demand be significantly higher than normal.

Figure 6.1 shows the stock balances generated by OPSCON's STOCKMAN file LEVELPOL using the options VIEWGRAPHS and INITIAL. This shows a simulation of stock balances for an initial three

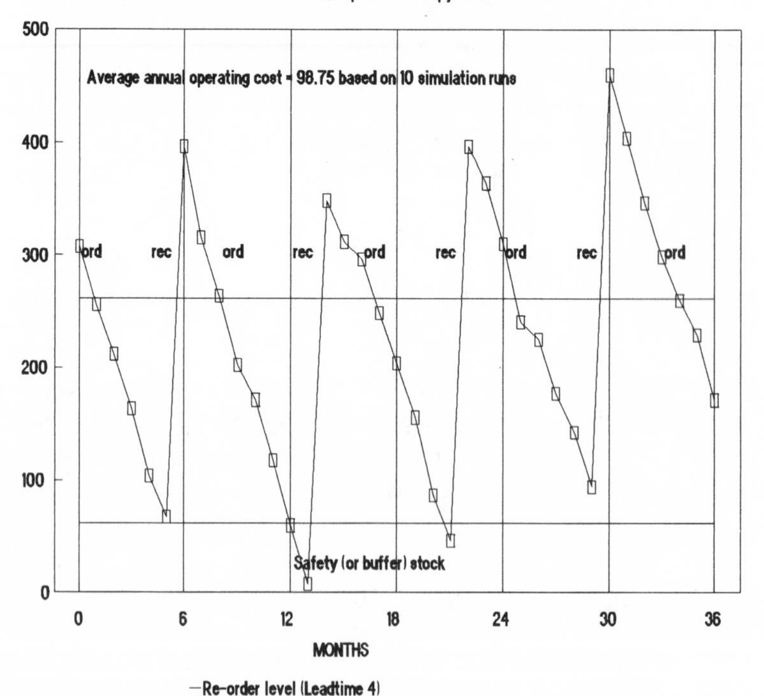

6.1 Re-order level policy stock balances for initial three year period.

year period (assuming monthly demand) and indicates when orders for replenishment are raised (ord) – as a result of the re-order level being equalled or broken – and subsequently received (rec) after the leadtime delay, which in this particular situation is set at four months.

For the initial three year period shown here, a reasonable level of service appears to be offered with the re-order level set at 260 units since, because demand during the supplier leadtime never exceeds the value of re-order level, no stockouts (i.e. a failure to meet customer demand) occur and customers of the stock system never have to wait for their demands to be met. Figure 6.2 (generated using the TOTAL option within VIEW-GRAPHS) shows a less detailed view of the same situation but taken over a longer, twelve year, period. This also confirms that a re-order level of 260 units appears to offer a reasonably high level of service since only six stockouts occurred in the twelve year period. For this situation the average annual operating cost, based on ten simulation runs of the inventory control system, is shown to be 98.75. The operating cost includes:

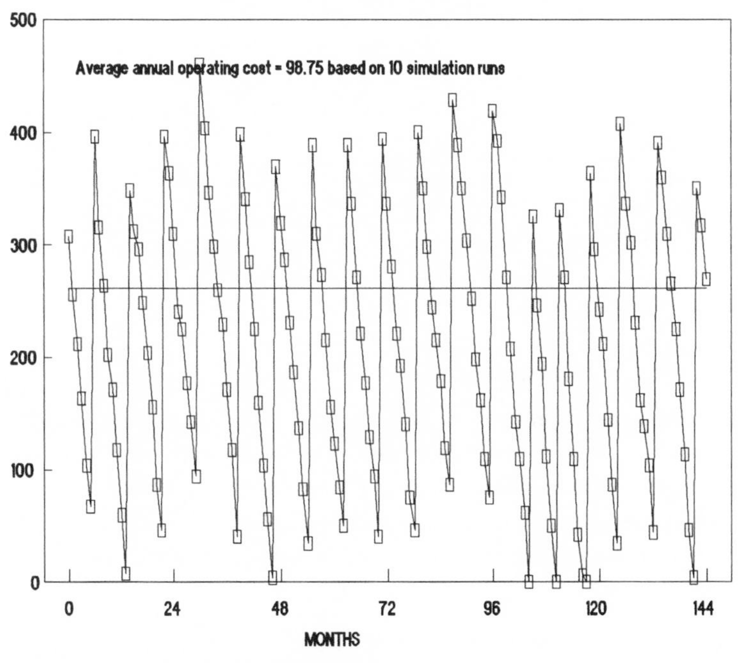

6.2 Re-order level policy stock balances for initial twelve year period.

- storage costs incurred through the stockholding process;
- ordering costs incurred as a result of placing replenishment orders; and
- penalty costs incurred during those periods when stockouts occur.

Examination of either or both of Fig. 6.1 and 6.2 indicates that the operation of the re-order level inventory control policy is determined completely by the setting of just two parameters, namely the value of the re-order level and the value of the replenishment quantity.

The traditional approach to establishing the value of the re-order level

To establish an appropriate value of the re-order level it is necessary to estimate or assume values for:

- $\bar{D}$ – the average demand per unit of time (usually estimated from the forecast). The higher the average demand, the more stock that will be required to meet the average demand during the supplier leadtime.
- σ_d – the standard deviation of demand per unit of time which for short-term forecasting models would usually be equated as the Mean Absolute Deviation of forecasting errors multiplied by 1.25 but which for medium-term forecasting models would normally be evaluated as the square-root of the Mean Squared Error. A higher variability of demand (and hence a larger value of the standard deviation of demand during the leadtime) requires more stock to absorb those variations if a reasonable level of service is to be achieved.
- L – the duration of the procurement leadtime, measured in the same periods or units of time as that on which the average and standard deviation of demand are based (within STOCKMAN time units can usually be defined as weeks or months). Since the re-order level triggers the placing of a replenishment order, the resulting leadtime delay in that order being received is the period of time when the stock control policy is at risk of running out, irrespective of the size of that replenishment order. The duration of the supplier leadtime is usually assumed, within inventory control theory, to be fixed in value to avoid mathematical complications: even though in practice it is recognised that two leadtimes are rarely the same.

In addition to the above, it is necessary to make an assumption as to the likely probability distribution of demand during the leadtime. With regard to this latter point, for reasonably fast moving items with an average

```
C.A.L. Module              OPERATIONS CONTROL - Stock Control                    CDL/96
PARAMETERS DEMAND COSTS SIMULATE INFORMATION/PRINT VIEWGRAPHS colOUR QUIT
Specify values of re-order level and replenishment qty

        STOCK CONTROL SIMULATION   :   RE-ORDER LEVEL POLICY  12 years
        ====================================================
   COSTS:                                PARAMETERS:        Actual Theoretical
     Storage(%pa mat'l cost)   22.50%  Re-order level..     260         260
     Material(+lab +o'heads)    1.00   Replen. qty.....     400         400
     Ordering cost/occasion    30.00   Vendor service..    87.5%       97.7%
     Penalty cost/period       30.00   Customer service    99.0%       99.9%
   DEMAND/ MONTH   Specified  Actual SIMULATED   COSTS/ANNUM
     Fcst average      50.00   51.34   Storage costs...........        46.88
     Fcst std dev      15.00   15.49   Ordering costs.........         45.00
     Leadtime (1 to 8)......       4   Penalty costs...........         7.50
   STOCKTURN...............     2.98   Total operating costs....       99.38
   ==========================================================================
   Months  Previous Current Received Current  Backorder O'shoot   Stockout
             stock   demand   orders   stock      qty.   amount     occur.
   ==========================================================================
        0     364       28               336
        1     336       52               284
        2     284       70               214                46
        3     214       44               170
File:LEVEL                      Esc toggles menu                                  MENU
```

6.3 STOCKMAN's LEVELPOL screen display showing information relating to demand situation where average demand $\bar{D} = 50$ and standard deviation of demand $\sigma_d = 15$ with a supplier leadtime of 4. The actual re-order level controlling the simulation is set equal to the theoretical value of 260 computed using Equation [6.1].

demand of twenty or more per period (time unit) it is often reasonable to assume that the demand during the leadtime is likely to be distributed normally.

The screen display (generated by the file LEVELPOL) associated with the stock balances shown in Fig. 6.1 and 6.2 (where clearly a re-order level value of 260 units appears to offer a reasonable level of service to customers at a reasonable cost) is displayed as Fig. 6.3. Examining the detail of this screen shows that the stock balances have resulted from a situation where:

- the demand per unit time was assumed to be distributed normally with an average $\bar{D} = 50$ units/month, a standard deviation $\sigma_d = 15$ units/month; and
- the duration of the leadtime delay in obtaining replenishments was fixed at $L = 4$ months.

It was then assumed that the demand during the leadtime would also be distributed normally with:

- an average of $\bar{D}L = 50 \times 4 = 200$ units/leadtime; and
- a standard deviation of $\sigma_d\sqrt{L} = 15\sqrt{4} = 30$ units/leadtime.

NOTE: In statistics, since the variance of a sample varies with the size of the sample, the standard deviation - which is the square-root of the variance - varies with the square-root of the sample size. In this case because the demand during the supplier leadtime is being estimated, the leadtime duration is effectively the sample size and $\sigma_d\sqrt{L}$ is therefore the best estimated value of the standard deviation of demand during the leadtime.

Based on these assumptions, the average demand during the leadtime is clearly $\bar{D}L$. If the re-order level were to be set at a value equal to the average demand during the supplier leadtime, since the probability of exceeding the average is 50%, the probability of a subsequent stockout occurring would also be 50%.

Clearly a 50% probability of running out of stock subsequent to placing a replenishment order is unlikely to offer an acceptable level of service to any inventory control system's customers. Hence, in practice, to offer a satisfactory level of service, the re-order level needs to be set at a value greater than the average demand during the leadtime. To establish the maximum expected demand during the supplier leadtime, rather than the average, the probability of that value not being exceeded could be specified. Assuming that a probability of 2.275% were agreed, the re-order level could then be set equal to the maximum expected demand during the leadtime using Equation [6.1]:

$$M = \bar{D}L + u_{0.02275}\sigma_d\sqrt{L} \qquad [6.1]$$

where $u_{0.02275} = 2$ is the value of the normal standard variable whose probability of being exceeded is 0.02275 or 2.3% (see Appendix C).

For the situation under consideration this would lead to:

$$M = 50 \times 4 + 2 \times 15\sqrt{4} = 260 \text{ units}$$

This relationship can be confirmed pictorially by referring to Fig. 6.4.

The relatively simple statistical relationship just described has been used traditionally to establish that if an order for replenishment were to be raised when the stock-on-hand was *exactly* equal to the re-order level, then in theory on 97.7% of these occasions when the inventory control system was at risk of a stockout (i.e. subsequent to a replenishment order being placed) demand during the leadtime would be less than the value of the re-order level and no stockout would occur. Conversely, on 2.3% of those occasions, demand during the leadtime would exceed the value of the re-order level and a stockout would occur.

Therefore, on the assumption that demand during the leadtime could be assumed to be normally distributed, this procedure establishes the value of the re-order level by defining a 'vendor service level' as being the

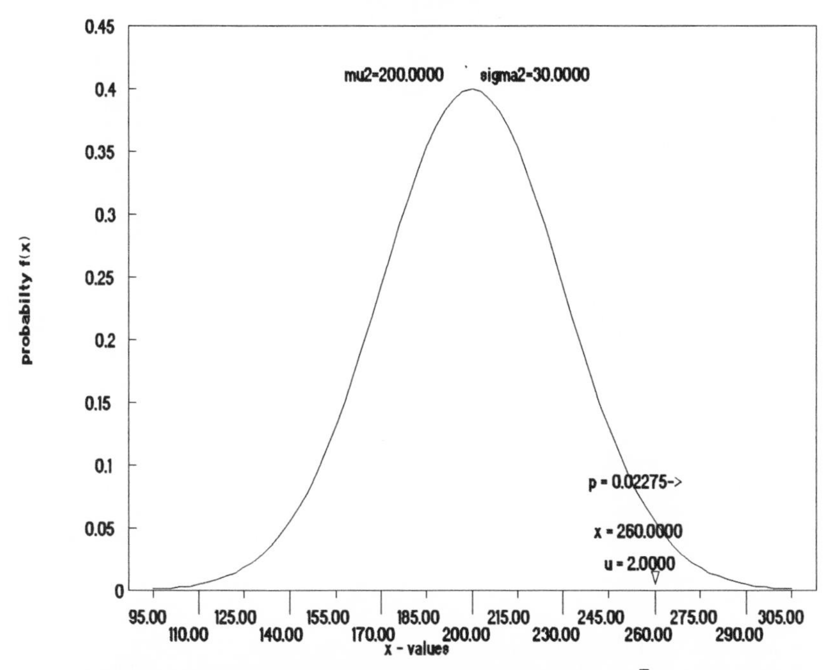

6.4 Normal distribution indicating with mean mu2 set at $\bar{D}L = 200$ and the standard deviation sigma2 set at $\sigma_d\sqrt{L} = 30$ that the probability of exceeding 260 is 0.02275 or 2.3%. Hence probability of not running out is equal to 97.7%.

probability of not running out of stock on those occasions the policy was at risk of running out, i.e. the probability of a stockout not occurring subsequent to an order for replenishment being raised but prior to its being received. This traditional approach of setting the re-order level value within a fixed order quantity system by defining a 'vendor service level' can be criticised on several counts (see Chapter 8). However, in practice, it still remains a popular method since, in spite of its flaws, the approach still achieves reasonably acceptable results.

Readers wishing to confirm that the demand pattern used in the STOCKMAN's LEVELPOL file's simulation is reasonably normal can use the VIEWGRAPHS option followed by the DEMAND and FREQUENCY options to generate Fig. 6.5 and 6.6 respectively. The former shows a scatter plot of the demand pattern and the latter a vertical bar chart of the frequency of occurrence of monthly demand values which

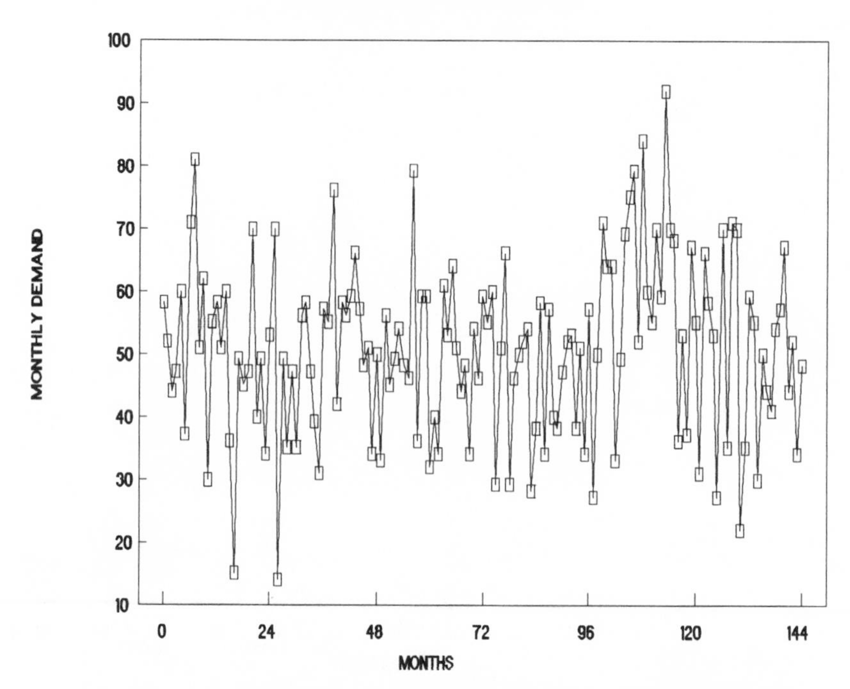

6.5 Pattern of monthly demand imposed on the stock control system with an average demand $\bar{D} = 50$ and a standard deviation of demand $\sigma_d = 15$.

exhibits the characteristic pattern associated with the normal distribution. A specified mean $\bar{D}$ of 50 per month and a standard deviation σ_d of 15 per month confirms, as one would expect, that the vast majority of values fall within the range $\bar{D} - 2.5\sigma_d$ and $\bar{D} + 2.5\sigma_d$, i.e. from 12.5 to 87.5 units per month.

Task 6.1. Using STOCKMAN's LEVELPOL file examine the practical range of the re-order level values for the current demand and supplier leadtime using the PARAMETER option followed by the LEVEL option. What assumptions has the package made in determining the maximum and minimum values of the range? Within the practical range of the re-order level:

1. examine the operating costs and establish where costs are minimised and comment on the reasons why operating costs increase at either end of the range; and
2. record the values of the theoretical and actual vendor service levels and comment on how these vary with the re-order level settings.

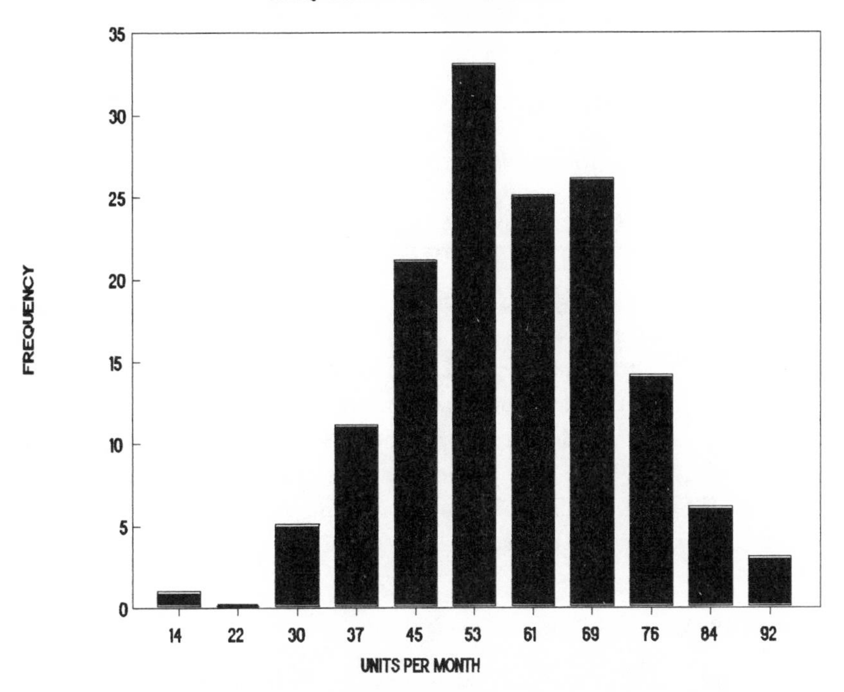

6.6 Frequency distribution of monthly demand with an average demand $\bar{D} = 50$ and a standard deviation of demand $\sigma_d = 15$.

Linking the forecast of demand to the setting of the re-order level

Within the re-order level policy simulation file LEVELPOL it is possible to change the average and standard deviation of demand to any set of chosen values. However, in a practical inventory control environment, these values would be established as a result of an earlier analysis of some real demand data with a forecasting package. To mimic this facility within OPSCON, when quitting any of FOREMAN's forecasting files the user is offered the option of exporting the results of the forecasting analysis to a temporary file for subsequent importation to either of the main inventory control files LEVELPOL or CYCLEPOL (see Chapter 9). When quitting in particular from FOREMAN files whose forecasting models assume growth or seasonality, because the forecast ahead varies over the forecast

Task 6.2. Initially use any of FOREMAN's forecasting files to analyse either demand data provided by the appropriate DEMONSTRATION option or demand data previously saved as an ENTIRE file using the DATAHELP facility. When using the QUIT option to terminate forecasting analysis, use the subsequent EXPORT option to retain the results of the analysis before proceeding automatically to STOCKMAN.

Subsequently, load STOCKMAN's LEVELPOL file and use the DEMAND and FORECAST options to load the retained forecast data which effectively becomes the average demand ($\bar{D}$) and the standard deviation of demand (σ_d). Confirm that the re-order level value evaluated is appropriate for the original demand data and justify why.

horizon, the user is asked to specify from which period ahead the forecast value is to be taken.

Practitioners with a real stocked item to consider should recognise that when the forecast data is imported into inventory policy simulation files this does not include any cost data. To establish a realistic cost environment it will at least be necessary to change the material costs via the COSTS and MATERIAL options; the storage and ordering costs default values being a realistic 22.5% and 30% respectively.

Conclusion

The setting of a re-order level to trigger the placing of replenishment orders is one of the most popular methods of raising orders to replenish an inventory system. The approach has an intuitive appeal and, when the item being stocked is physically small or a powder or liquid, a re-order level policy can then be implemented as a so called two bin system within which when a first bin is emptied, material is then withdrawn from a second bin whose capacity is equal to the re-order level. In this chapter the calculation of the correct re-order level in line with demand information and the length of the supplier leadtime has been considered.

File from OPSCON's package STOCKMAN associated with this chapter

LEVELPOL file – a complex, multi-level menu driven file which simulates the stock balance of the re-order level policy.

The controlling parameters of the policy, namely the re-order level and the replenishment quantity are evaluated in the light of demand and costs conditions and the user can either use the recommended values or specify alternative values.

The demand environment can be specified in terms of the average and standard deviation of demand per unit time and the duration of the replenishment leadtime. Alternatively the demand environment can be established as a result of a previous forecasting analysis from which the relevant forecasting information can be imported.

The file contains a comprehensive cost model and users can examine the results of the simulation model through a variety of graphs which display:

- initial and overall stock balances together with detail of stockouts where these occur;
- overshoot of the re-order level;
- the demand environment – in terms of a scatter plot of demand orders and a frequency distribution of those orders; and
- a breakdown of storage, ordering, penalty and purchase/acquisition costs.

All simulations can be run under the assumptions that backordering (the assumption that demand orders that cannot be met immediately are nevertheless accepted thus inferring negative stock) can either be allowed or prohibited. To provide relatively stable cost analyses, when a simulation is run the average costs resulting from ten simulation runs are reported.

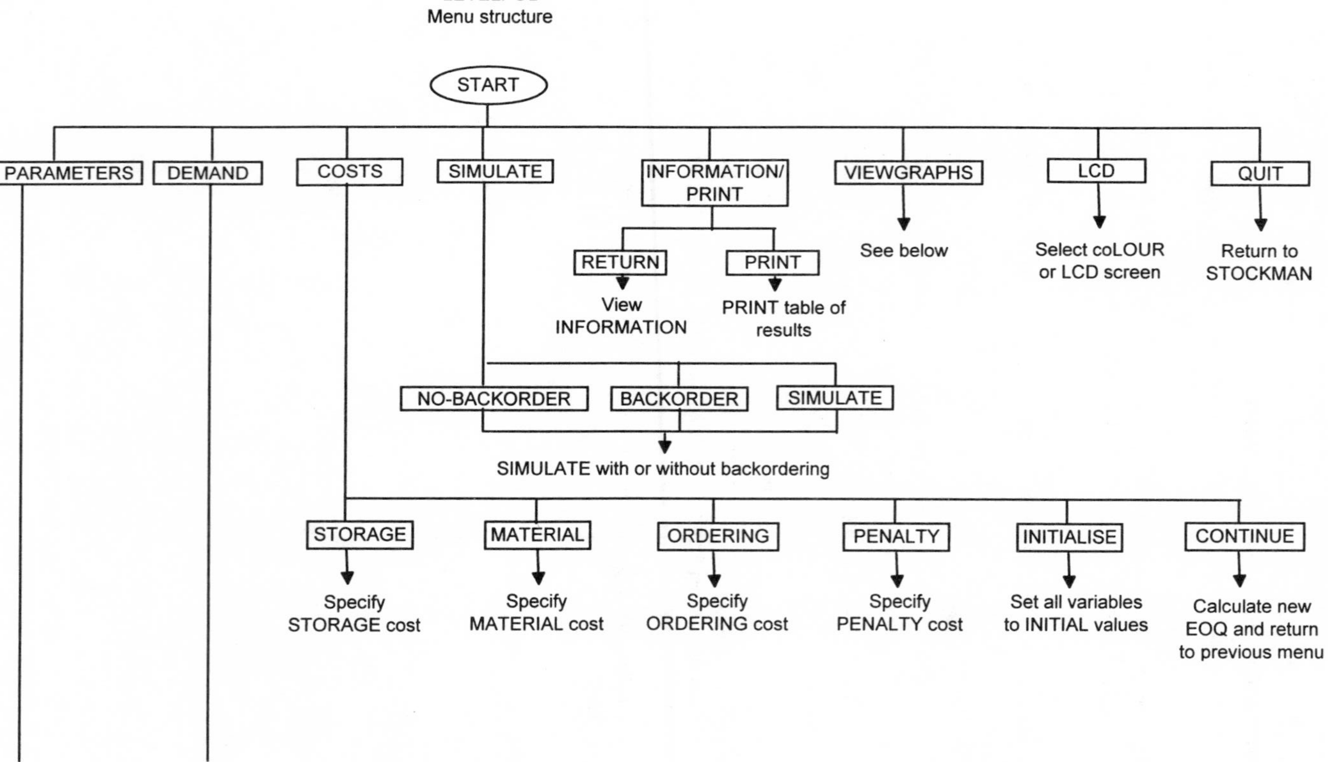
LEVELPOL
Menu structure
START
PARAMETERS
DEMAND
COSTS
SIMULATE
INFORMATION/ PRINT
VIEWGRAPHS
LCD
QUIT
See below
Select coLOUR or LCD screen
Return to STOCKMAN
RETURN
PRINT
View INFORMATION
PRINT table of results
NO-BACKORDER
BACKORDER
SIMULATE
SIMULATE with or without backordering
STORAGE
MATERIAL
ORDERING
PENALTY
INITIALISE
CONTINUE
Specify STORAGE cost
Specify MATERIAL cost
Specify ORDERING cost
Specify PENALTY cost
Set all variables to INITIAL values
Calculate new EOQ and return to previous menu

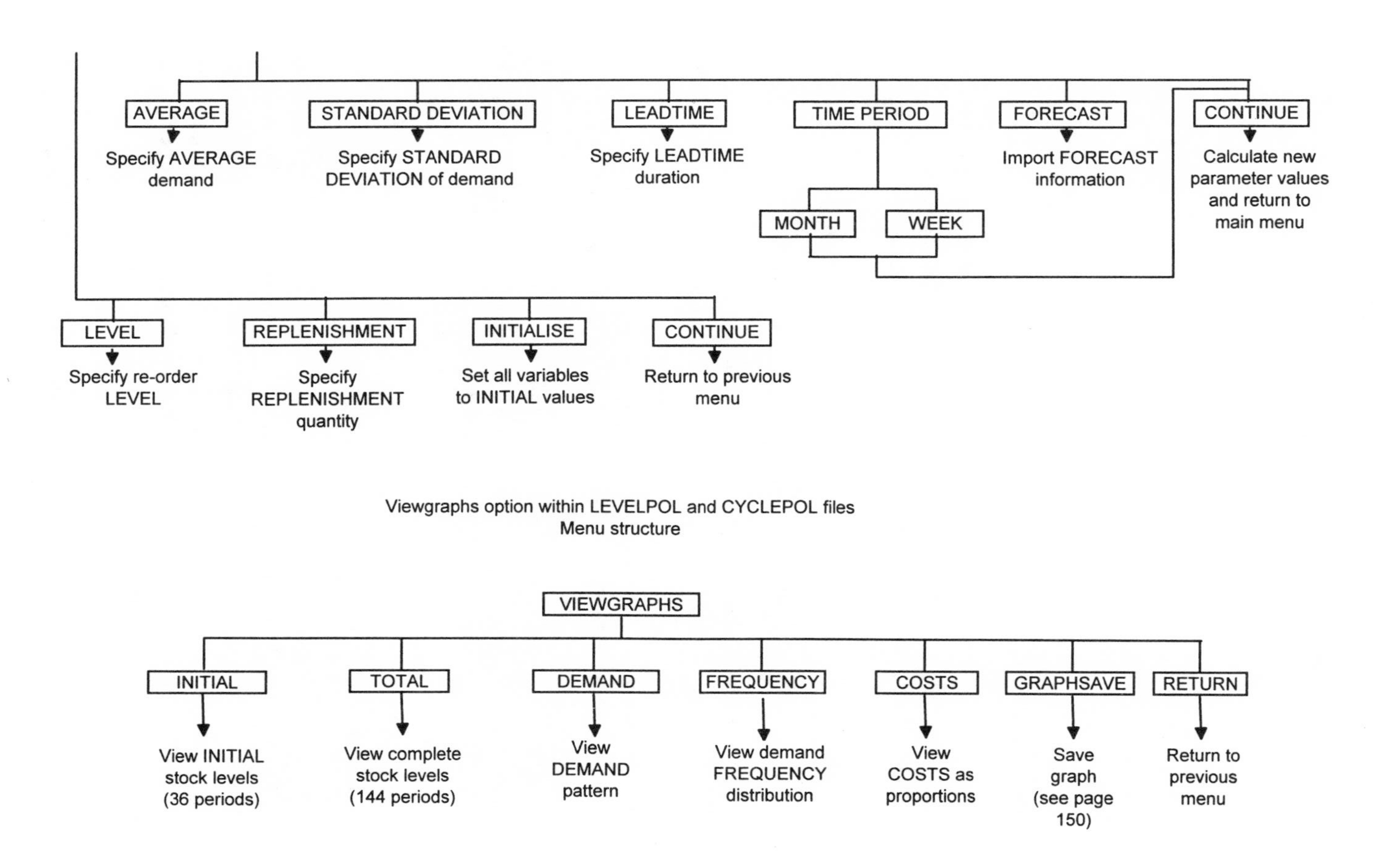

Viewgraphs option within LEVELPOL and CYCLEPOL files
Menu structure

7

Establishing the size of the replenishment order

Introduction

Having established in Chapter 6 that the value of the re-order level within a re-order level inventory control policy could be based solely on the concept of offering a certain level of service; traditionally the size of the replenishment order has then been determined using a totally unrelated approach of minimising the annual cost of operating that inventory policy. This produces a value for the replenishment quantity which is often referred to as the Economic Order Quantity (EOQ).

As with the method of establishing the value of the re-order level, this EOQ method of establishing the value of the replenishment quantity also has many critics. However, the approach is both well known and widely used and has yet to be supplanted by an alternative approach which has as wide an appeal. However, to place the EOQ in perspective, some of the criticisms of the approach will be addressed in this chapter. The particular criticism that the replenishment quantity should not be evaluated separately on a cost basis but should be considered jointly with the evaluation of the re-order level is considered in Chapter 8.

The EOQ approach to establishing the value of the replenishment quantity

The theory behind establishing the value of the replenishment quantity has been generally attributed to Wilson[13] who, as early as 1934, proposed evaluating the size, q, of the replenishment order on a basis of minimising the annual cost of operating an inventory policy, this cost being assumed to be a combination of the annual cost of ordering replenishment orders and the cost of storage per annum.

The underlying assumptions of this approach are that:

- the ordering cost (i.e. cost of raising replenishment orders) per annum is assumed to be equal to the number of replenishment orders multiplied by the cost C_o of raising each order. Establishing from a forecasting analysis that the annual demand is estimated as A, then the annual cost of raising A/q replenishment orders per annum is given as:

$$\left(\frac{A}{q}\right) C_o$$

where, for monthly demand estimated at $\bar{D}$, the annual demand $A = 12\bar{D}$ but for weekly demand $A = 50\bar{D}$;

- the storage cost per annum is assumed to be based on the average level of stock held, this latter being evaluated as half the replenishment order size q (i.e. q/2) plus a fixed amount of safety stock (S) which it is assumed is independent of replenishment order size. Given that a stocked item's price (unit material cost) is C_m and that the percentage cost of holding stock per annum is given as i, then it follows that the cost of storage per annum is equal to the average stock $(q/2 + S)$ multiplied by the unit cost of storage per annum iC_m and is, therefore, evaluated as:

$$\left(\frac{q}{2} + S\right) iC_m$$

On the basis that ordering costs decrease parabolically with replenishment order size but that storage costs increase linearly, it then follows that if the annual cost, C, of operating an inventory policy is assumed to be equal to the cost of ordering replenishments plus the cost of storage then:

$$C = \left(\frac{A}{q}\right)C_o + \left(\frac{q}{2} + S\right)iC_m \qquad [7.1]$$

By differentiating the annual operating costs, C, (as defined by Equation [7.1]) with respect to the replenishment order quantity, q, produces:

$$\frac{dC}{dq} = \frac{-AC_o}{q^2} + \frac{iC_m}{2} \qquad [7.2]$$

It is then assumed that the value of the replenishment order which produces the minimum annual operating cost is defined as the order size corresponding to that position where the tangent to the total operating cost curve is zero, i.e. $dC/dq = 0$. This results in a value for the replenishment quantity, referred to as the Economic Order Quantity, Q_o, which is calculated by the well known square-root formula developed as the solution to Equation [7.2] when $dC/dq = 0$, and produces:

$$Q_o = \sqrt{\frac{2AC_o}{iC_m}} \qquad [7.3]$$

Readers unfamiliar with calculus can rest assured that these results are valid since the Economic Order Quantity is possibly the most publicly debated topic in management science.

To illustrate the use of the EOQ equation, the initial stock control situation depicted earlier in Chapter 7 is defined by the following input information:

A (annual demand) = $12 \times \bar{D} = 600$ with $\bar{D} = 50$ units per month,
C_o (the cost of raising an order per occasion) = 30,
i (the annual holding interest rate) = 22.5% (i.e. 0.225); and
C_m (the stocked item's unit value) = 1.00.

By substituting in Equation [7.3], the Economic Order Quantity Q_o is evaluated as:

$$Q_o = \sqrt{\frac{2 \times 600 \times 30}{(0.225 \times 1)}} = 400 \text{ units}$$

The calculations above can be confirmed by examining Fig. 7.1 which is the main screen display of STOCKMAN's EOQ_EVAL file.

```
C.A.L. Module            OPERATIONS CONTROL - Stock Control                  CDL/96
STORAGE  MATERIAL   ORDERING   DEMAND   VIEWGRAPHS   QUIT
Specify storage cost as percentage p.a.

          *** INTRODUCTION TO ECONOMIC ORDER QUANTITY ***
          ===============================================
  INPUT COSTS:                      DEMAND DATA:
   i: Storage p.a.     (%Cm) 22.50% D: average demand per month        50
   Cm: Material(lab+o'heads)  1.00  A: annual  demand (12*D)           600
   Co: Ordering/occasion     30.00  ACm: Purchase/acqu. costs p.a.     600

  ECONOMIC ORDER QUANTITY:       Qo..   400 units
  Qo = √((2)(A)(Co)/(i)(Cm))    =    √((2)(600)(30)/(0.225)(1.00))     = 400
   Total operating costs p.a. at Qo (ordering cost = storage cost)
   C(Qo)=(A/Qo)Co+(Qo/2)(i)(Cm)=(600/400)30+(400/2)(0.225)(1.00) = 90.00
                              =        (45.00)  +  (45.00)           = 90.00
   Differential of annual operating costs at Qo (zero for minimum cost)
   dC/dq=-A(Co)/(Qo^2)+(i)(Cm)/2 = -600(30.00)/(400^2)+(0.225)(1.00)/2=0

  =====================================================================
  ACTION: Alter input costs and monthly demand and observe effect on
          size of the Economic Order Quantity  -  Qo
    NOTE: Irrespective of input costs; ordering costs = storage costs
          and differential of cost curve dC/dq = 0 for minimum cost
File:EOQ_EVAL                  Press Esc if no menu                         MENU
```

7.1 STOCKMAN's EOQ_EVAL files screen showing both the formulation of the equations for Q_O, C and dC/dq together with the resulting calculations in line with the input costs and demand data.

Task 7.1. Using STOCKMAN's EOQ_EVAL file as a NEW USER, develop a better understanding of the development of the economic order quantity, Q_O, and confirm that at the EOQ:

- the cost of ordering will always be equal to the cost of storage; and;
- the differential of the cost equation dC/dq is always zero.

Subsequently use the file to examine how the economic order quantity varies with:

- changes in annual demand, A;
- changes in ordering costs, C_O;
- changes in material costs, C_m; and
- changes in storage costs, i.

This file:

- demonstrates that irrespective of the value of the Economic Order Quantity, at the EOQ the cost of ordering will always be equal to the cost of storage; and
- confirms that irrespective of the cost and demand structure used to formulate the EOQ, at the EOQ the differential of the cost equation dC/dq is always zero.

Readers having used STOCKMAN's EOQ_EVAL file will also have noted from Fig. 7.2 that:

- the cost of ordering can be seen to fall rapidly as the size of replenishment order increases since fewer, larger orders are needed to meet the same annual demand;
- the cost of storage can be seen to rise linearly with replenishment order size since the average stock held also rises with increased replenishment order size; and
- the annual operating cost, equal to the sum of the ordering and storage costs, can be seen to be a minimum at the point at which the costs of ordering and storage intersect and where in addition the differential to the operating cost curve $dC/dQ = 0$.

The other important relationship demonstrated by the EOQ_EVAL file is the relative size of the operating cost (ordering + storage costs) compared with the purchase/acquisition cost. More specifically, Fig. 7.3 provides:

- further confirmation that at the EOQ ordering cost and storage cost are equal; and
- confirmation of the relatively small size of the operating cost (i.e. ordering and storage) relative to the purchase/acquisition cost.

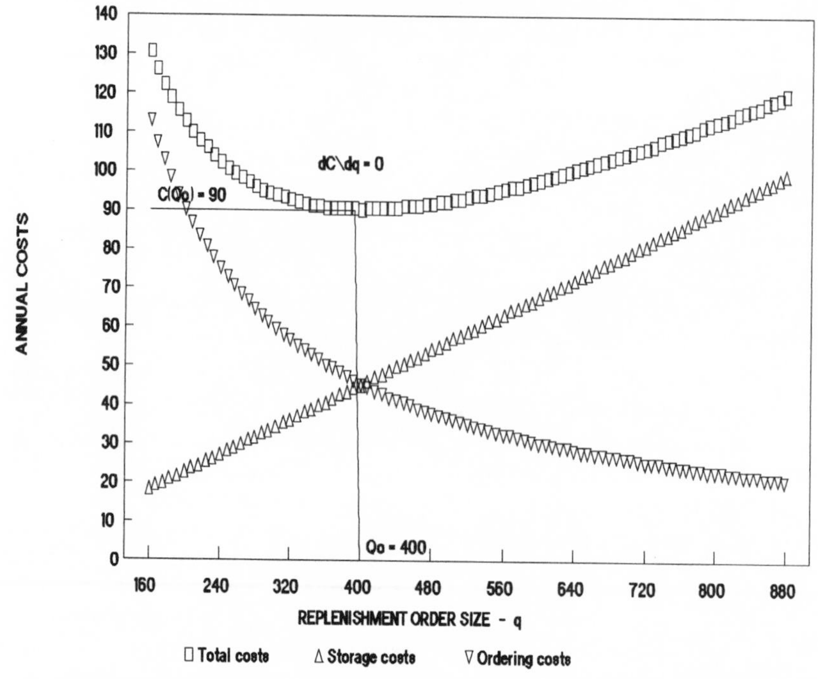

7.2 Annual ordering, storage and total operating costs versus replenishment order size.

Criticisms of the EOQ approach

There are many criticisms of the simple EOQ approach some of which are:

- that the costs involved in the formulation of the EOQ are difficult to estimate with any great accuracy, and that the storage cost in particular may not increase linearly with replenishment order size but incrementally as storage capacity limits are reached;
- that rather than minimising costs the EOQ should be formulated on a criterion of maximising profit, return on investment, etc;
- the total operating cost curve is so shallow it makes little difference whether the chosen replenishment quantity is near to the EOQ or not; and
- the EOQ approach does not take into account the possibility of achieving price breaks by placing large replenishment orders on suppliers.

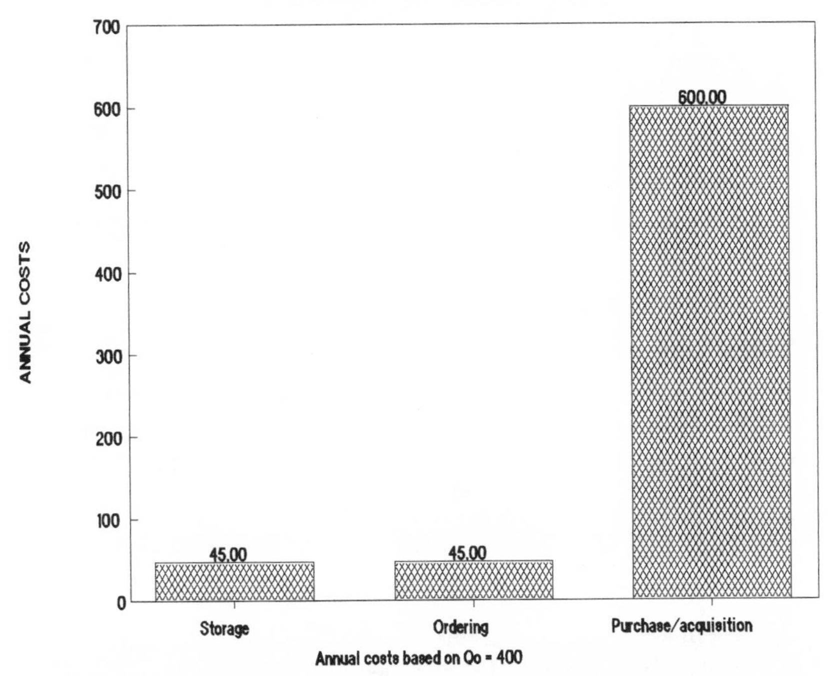

7.3 Bar chart showing the equal size of the ordering and storage costs at the Economic Order Quantity and the relative size of these costs compared with the purchase/acquisition cost.

These latter two criticisms will now be examined in detail.

Cost sensitivity of the EOQ approach in establishing the replenishment order size

In general it can be seen from Fig. 7.2 that if a replenishment quantity other than the EOQ were to be used, since storage costs increase linearly above the EOQ whereas ordering costs increase more steeply below the EOQ, then proportionately it would be preferable to place a replenishment order larger than the EOQ rather than smaller. This general observation can be confirmed algebraically by considering the general case of replenishment order, q, whose size is x% larger or smaller then the EOQ, Q_o.

Given that:

$$q = Q_o\left(1 \pm \frac{x}{100}\right)$$

It follows that if safety stock, S, can be ignored since it is independent of replenishment order size, then the annual operating cost, C_q of operating at q rather than at Q_o will be given by transposing Equation [7.1] such that:

$$C_q = \frac{AC_o}{Q_o}\left(1 \pm \frac{x}{100}\right) + Q_o\left(1 \pm \frac{x}{100}\right)\frac{iC_m}{2} \qquad [7.4]$$

and that the cost, C_{Q_o} of operating at Q_o will be given as:

$$C_{Q_o} = \frac{AC_o}{Q_o} + \frac{Q_o iC_m}{2} \qquad [7.5]$$

simplifying and combining the relationships expressed in Equations [7.4] and [7.5], it can be shown that the percentage increase in annual operating costs y% resulting from using a replenishment order size x% higher or lower than the EOQ is given by:

$$y = 100\left(\frac{C_q}{C_{Q_o}} - 1\right) = \frac{x^2}{2(100 + x)} \qquad [7.6]$$

By defining a range of replenishment order sizes x% higher or lower than the EOQ, the relationship expressed in Equation [7.6] can be seen in detail in Table 7.1.

Examination of Table 7.1 confirms the fact that replenishment quantities larger than the EOQ involve proportionately lower increases in the annual operating cost than smaller quantities. Hence, a replenishment order size 50% larger than the EOQ causes an 8.33% increase in operating cost whereas an order size 50% smaller causes an increase of 25%. Table 7.1 also demonstrates that the rate of increase in operating cost in both situations is considerably smaller than the imposed proportionate increase or decrease in the size of the replenishment order.

Table 7.1 Detailing the percentage increase in annual operating costs when specifying a replenishment quantity, q, chosen to be larger or smaller than the EOQ, Q_o.

Percentage increase or decrease in replenishment quantity, q, as compared with the EOQ, Q_o	x	−50	−25	0	+25	+50	+75	+100
Percentage increase in annual operating cost	y	25.00	4.17	0	2.50	8.33	16.07	25.00

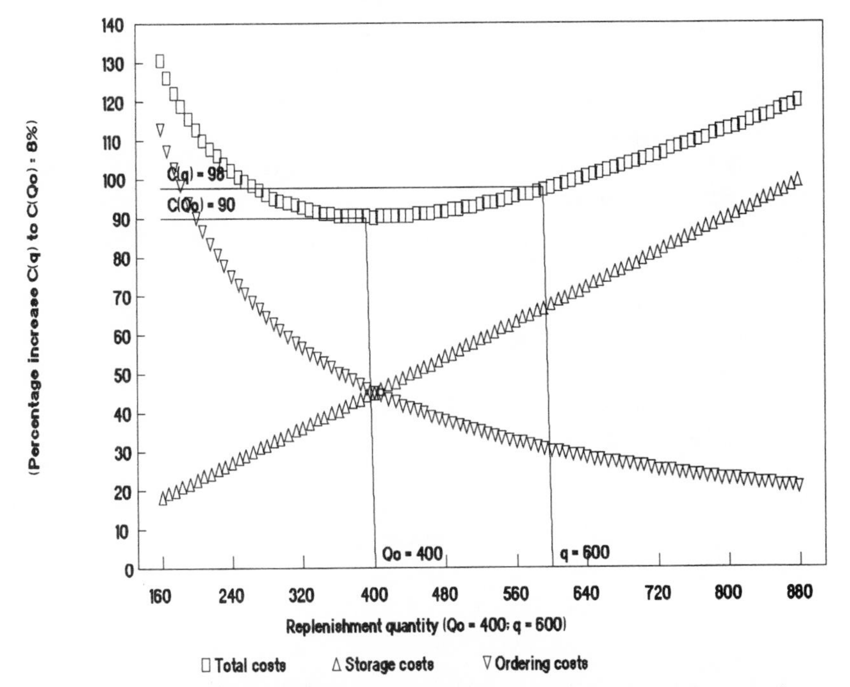

7.4 The position on the operating cost curve of a specified order quantity, q, relative to the Economic Order Quantity, Q_0.

Readers can confirm this relationship by using STOCKMAN's EOQ_SENS file which allows the user to specify replenishment order sizes other than the EOQ and to examine the subsequent costs incurred. Where the size of the chosen replenishment order is not too close to the EOQ, its position relative to the EOQ is also shown pictorially using the VIEWGRAPHS option – as shown in Fig. 7.4 where the 50% increase in order size Q_o (noted in the main sub-heading) implies an 8% increase in costs (noted in the Y-axis heading).

Task 7.2. Using STOCKMAN's EOQ_SENS file, specify order sizes either side of the EOQ and confirm that the annual operating cost increases proportionately more steeply below the EOQ than above it.

What additional problems can occur when operating a re-order level inventory policy if the replenishment order selected is too small?

Discount required on purchase price to reduce the annual total cost

As can be seen clearly in Fig. 7.2, the annual purchase/acquisition cost (i.e. the actual cost of purchasing the equivalent of a year's annual customer demand) is an order of magnitude higher than the annual operating cost (i.e. the combined cost of ordering replenishments and storing the item under consideration). In the previous section it was shown that when operating with a replenishment order size larger or smaller than the EOQ, a relatively small increase in annual operating cost occurred. However, if the choice of a replenishment order size larger than the EOQ were to be matched by a discount in unit purchase price then because the annual purchasing/acquisition cost (which is directly influenced by any potential price discount) is so much larger than the operating cost even a very small percentage price discount for larger replenishment order sizes can offset the proportionately small increase in operating cost that will be incurred by operating with the larger quantity.

In practice there are many situations where, by ordering a larger replenishment order than the recommended EOQ, it is possible to negotiate a discount on unit price. Such opportunities occur, for instance, when a package or container size is involved and a discount is offered in preference to splitting the package or container.

Given that the proposed replenishment quantity at which a discount is on offer is x% larger than the EOQ, it can be shown (see Lewis[5]) that the percentage discount δ_T required to achieve a lower total annual inventory cost (i.e. annual purchase/acquisition + ordering + storage costs) can be defined by:

$$\delta_T > \frac{C_o x^2}{(1 + x)(C_o(1 + x) + Q_o C_m)} \qquad [7.5]$$

which for an order size 50% larger than the EOQ and for values of C_o, Q_o and C_m quoted earlier would require that the discount on purchase price be greater than:

$$\delta_T > \frac{30(.5)^2}{1.5(30*1.5 + 400*1.00)} = 1.12\%$$

and for a range of replenishment order quantities larger than the EOQ would, for a specific stocked item, require the following (Table 7.2) relatively low price discounts to keep the total annual inventory cost at least equal to that incurred if one were using the EOQ, Q_o.

Table 7.2 Percentage discount on unit purchase price required to maintain total annual inventory cost at or below that incurred at the EOQ

Percentage increase in replenishment quantity q as compared with EOQ	x	25	50	75	100	150	200
Price discount required to match total annual inventory cost when C_m = 1.00, C_o = 30, i = 22.5%, A = 600 and Q_o = 400	δ_T	0.34	1.12	2.13	3.26	5.68	8.16

The potential for price discounts when deciding on a replenishment order size larger than the EOQ – as described above – has been incorporated into STOCKMAN's LEVELPOL file such that if a replenishment order size 25% larger than the EOQ is proposed, a potential discount on the purchase price is automatically offered. This feature is demonstrated in Fig. 7.5 which is the INFORMATION screen display for the file. By examining Fig. 7.5, it can be seen that the prompt requesting a possible price discount figure incorporates the minimum value δ_T that would cause an overall reduction in the total cost of running the inventory system, namely purchase/acquisition + ordering + storage costs. Figure 7.6 provides a bar chart showing the relative sizes of these costs for both the proposed replenishment quantity, q, with the specified price discount and the Economic Order Quantity, Q_o, with no price discount.

```
C.A.L. Module          OPERATIONS CONTROL - Stock Control                CDL/96
Discount from 0% to 20% available. At least 1.12% required to reduce total costs
1.12

    Summary statistics for RE-ORDER  Operating and purchase costs based
    LEVEL POLICY simulation          on indicated replenishment quantities
    STORAGE                                               C(q)      C(Qo)
    Average stock held        316.11   Storage            69.54     48.62
    Material cost (Cm)          1.00   Ordering           30.00     45.00
    Price discount             1.12%   Penalty             2.50      3.75
    Storage cost p.a (i)      22.50% Purchase/acquisit.  593.28    600.00
    Item storage cost (iCm)     0.22 TOTAL ANNUAL COST   695.32    697.37
    Annual storage cost        70.33
    ORDERING                         PARAMETERS         Actual Theoretical
    Total orders placed           12 Re-order level  M     260        260
    Average overshoot             21 Replenishment qty     600        400
    Orders placed p.a.          1.00
    Ordering cost/occ. (Co)    30.00 Stockturn                       1.90
    Annual ordering cost       30.00   Annual demand value (A)     600.58
    PENALTY                            Average investment          316.11
    No. periods of stockout        1
    Stockout periods p.a.       0.08 Service levels     Actual Theoretical
    Penalty/stockout period    30.00   Vendor   SL      91.67%     97.72%
    Annual penalty cost         2.50   Customer SL      99.92%     99.96%
  File:LEVELPOL                     Esc toggles menu                      EDIT
```

7.5 The INFORMATION screen display from STOCKMAN's LEVELPOL file indicating percentage price discount required to produce an overall saving on the total annual inventory cost.

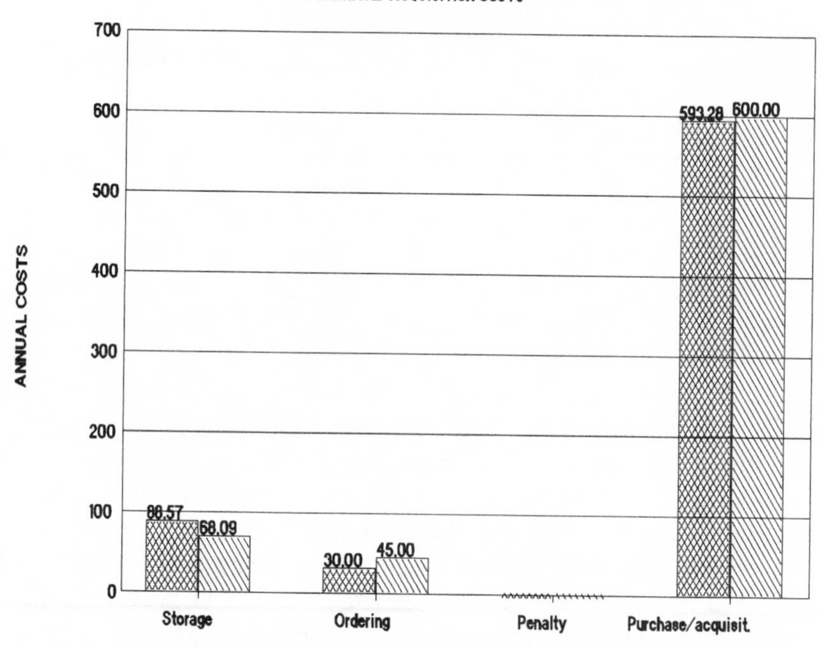

7.6 Bar chart of purchase/acquisition, ordering and storage costs confirming that, at the percentage price discount specified, an overall cost saving is made when operating with a replenishment quantity, q, and a price discount, δ_T.

Conclusion

The formulation of a replenishment order size known as the Economic Order Quantity is the most discussed single topic in management science generally and also perhaps the most criticised. This chapter, in describing the EOQ concept, has examined some of these criticisms. Other authors have confirmed that while the particular value of an EOQ might well be open to criticism, the overall concept does have considerable merit. In particular the approach known as coverage analysis proposes that the number of replenishment orders placed per annum, n, should be made proportional to the square-root of the annual usage value AC_m, i.e.

$$n = k\sqrt{AC_m}$$

where k is a constant.

Given an annual demand A being replenished by n orders of size q per year, this approach leads to a replenishment order size defined by:

$$q = A/n = A/\ k\sqrt{AC_m} = (1/k)\ \sqrt{A/C_m}$$

which if $k=\sqrt{i/2C_o}$ is simply another form of the Economic Order Quantity as originally defined in Equation [7.3].

Both coverage analysis and the original concept of the EOQ appear to produce replenishment order sizes which, in a multi-item inventory situation, creates a reasonably correct product mix in terms of relative order sizes. However, having established replenishment order sizes on an EOQ principle, it could then be argued that all such order sizes should be increased or decreased proportionately based on a further criterion such as minimising overall investment, meeting restrictions on storage space, etc.

Files from OPSCON's package STOCKMAN associated with this chapter

EOQ_ EVAL file – a file which for new users proceeds sequentially through the evaluation of the Economic Order Quantity (EOQ) based on an established cost and demand environment. Within this procedure:

- the formula for the EOQ together with the current value of the variables is developed and the user asked to indicate the value of the EOQ;
- the user is shown that, irrespective of the cost and demand environment, at the value of the EOQ the ordering costs always equal the storage costs; and
- the user is also shown that again, irrespective of the cost and demand environment, at the value of the EOQ the gradient to the cost curve is always zero – hence defining a minimum cost.

Subsequent to the above procedure the user is allowed to vary the cost and demand environment to examine the effect of these changes on the value of the EOQ. The graphs offered with this file are either a line graph of the operating (ordering and storage) costs or a bar chart of operating and purchase/acquisition costs.

EOQ_SENS file – similar in structure to the file EOQ_EVAL, this file offers the user the opportunity of specifying a replenishment order size other than the EOQ. The cost of operating at this replenishment order size can then be investigated and the sensitivity of the operating costs compared with those due to the EOQ established.

The graphs offered with this file are either a line graph of the operating (ordering and storage) costs showing the position of both the EOQ and the

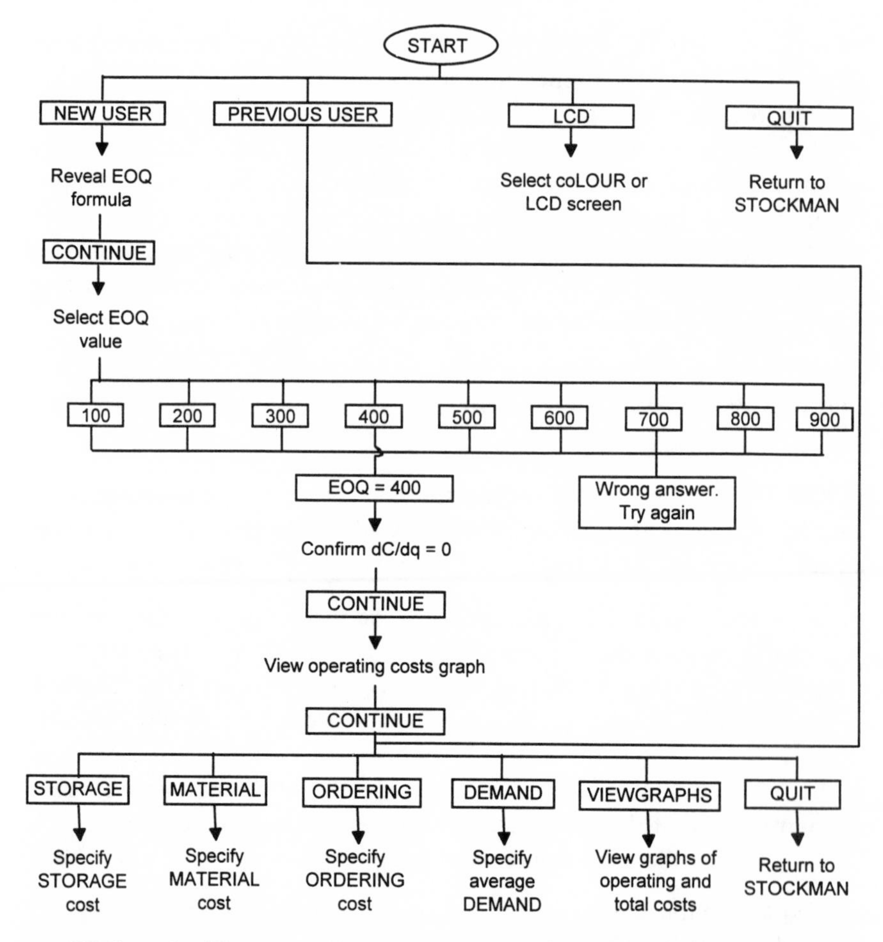

specified replenishment order quantity or a bar chart of operating and purchase/acquisition costs for both the EOQ and the specified replenishment order quantity.

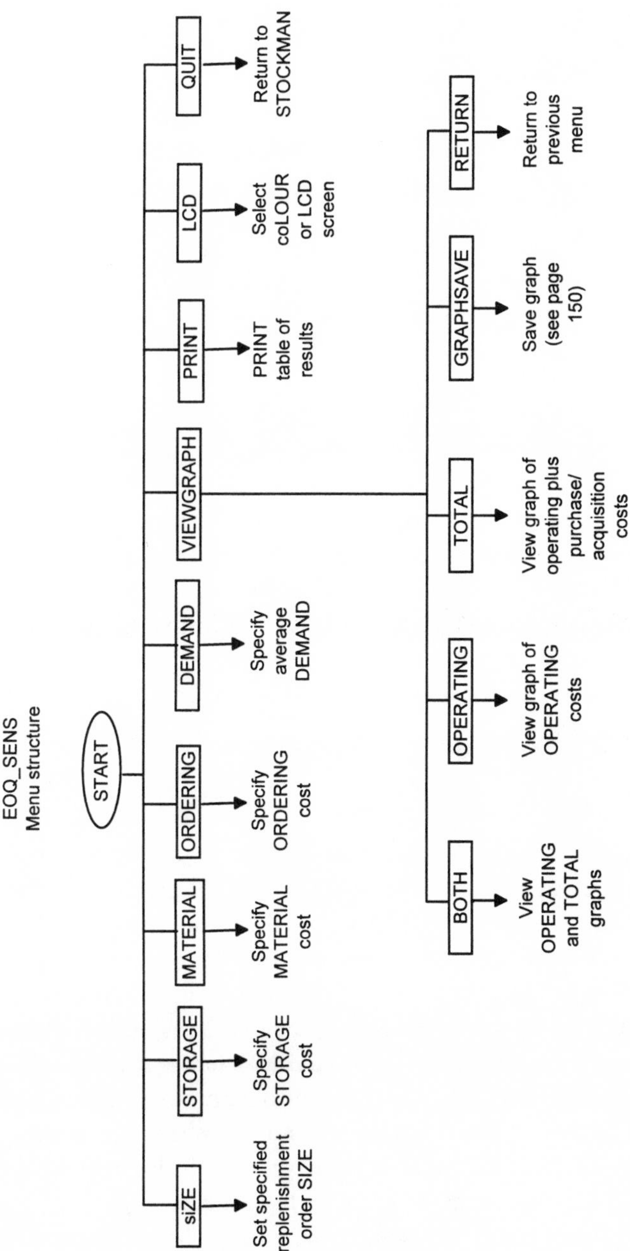
EOQ_SENS
Menu structure
START
siZE
Set specified replenishment order SIZE
STORAGE
Specify STORAGE cost
MATERIAL
Specify MATERIAL cost
ORDERING
Specify ORDERING cost
DEMAND
Specify average DEMAND
VIEWGRAPH
PRINT
PRINT table of results
LCD
Select coLOUR or LCD screen
QUIT
Return to STOCKMAN
BOTH
View OPERATING and TOTAL graphs
OPERATING
View graph of OPERATING costs
TOTAL
View graph of operating plus purchase/ acquisition costs
GRAPHSAVE
Save graph (see page 150)
RETURN
Return to previous menu

8

Examining the relationship between the re-order level and the replenishment quantity

Introduction

In discussing the setting of the two controlling parameters of the re-order level inventory policy, Chapter 6 considered the setting of the re-order level based on offering a specified level of service whereas Chapter 7 considered the evaluation of the replenishment quantity on a totally separate basis of minimising annual costs of ordering and storage.

In this chapter the relationship of these two controlling parameters is considered.

Interpreting the service level based on simple statistical concepts

For the demand situation considered in Chapter 6, p 86, traditional stock control packages would evaluate a re-order level of 260 units to offer a 97.7% vendor service level defined as the probability of not running out of stock on those occasions the policy was at risk of running out. As a totally separate exercise, for the costs and demand situation discussed in Chapter 7, p 96, the replenishment quantity would be calculated as the Economic Order Quantity of 400 units, i.e. that replenishment order size which minimised annual operating costs, i.e. the costs of storage and ordering.

Within the LEVELPOL file, whose main screen display was shown previously as Fig. 6.3 (see page 85), these two values appear as the theoretical re-order level and replenishment quantity. The corresponding actual values are the settings which are used to control the inventory control simulation process, that is the actual re-order level which triggers the placing of orders and the actual size of the replenishment quantity placed. These actual values can either be the same as the recommended theoretical values, as indicated in Fig. 6.3, or values set independently by the user.

Readers should note that although the re-order level value of 260 units is confirmed as offering a theoretical vendor service level of 97.7%, the resulting actual level measured within the simulation is always lower, as the value of 87.5% confirms in the situation shown in Fig. 6.3. Although some of this disparity can be put down to experimental errors in the simulation process, the main reason for this drop in service level has a specific cause which will be discussed later in this chapter.

Although the discussion describing the setting of the re-order level policy's controlling parameters in Chapters 6 and 7 appears logical and conclusive, and was indeed the basis for most stock control packages for many years, it subsequently became apparent that the underlying theory contained three fundamental flaws, namely:

1. The vendor service level – defined as the probability of not running out of stock on those occasions the policy was at risk of running out and evaluated using simple normal probability theory, although appearing to offer a logical approach to establishing the re-order level value, on closer inspection fails to identify either the frequency or severity of stockouts (i.e. how often stockouts occur or how many units of unfulfilled demand or stock shortage occur) or to indicate to customers of the stock control system what proportion of their demand would be met ex-stock (i.e. immediately). A very simple illustration of this feature would be to consider a probability of not running out of stock per occasion of 11/12 i.e. 91.66% which, alternatively, offers a probability of stockout of 1/12 i.e. 8.34%. At such a level of service, were replenishments to be ordered on a monthly basis clearly one stockout would occur per year on average. However with larger, quarterly replenishments a stockout would occur once every three years – even though in both cases the vendor service level would be 91.66! Hardly a convincing measure of service for customers.
2. The underlying formula used to calculate the re-order level assumes that when a replenishment order is raised the stock-on-hand *exactly equals* the re-order level. However, in practice this rarely occurs since in most situations the stock-on-hand will be below the re-order level at the instant the replenishment order is raised by an amount known as the overshoot. This feature can be confirmed using STOCKMAN's LEVELPOL file and using the VIEWGRAPHS and OVERSHOOT options, Fig. 8.1. This displays several stock balances in the vicinity of the re-order level together with the amount of overshoot which occurs

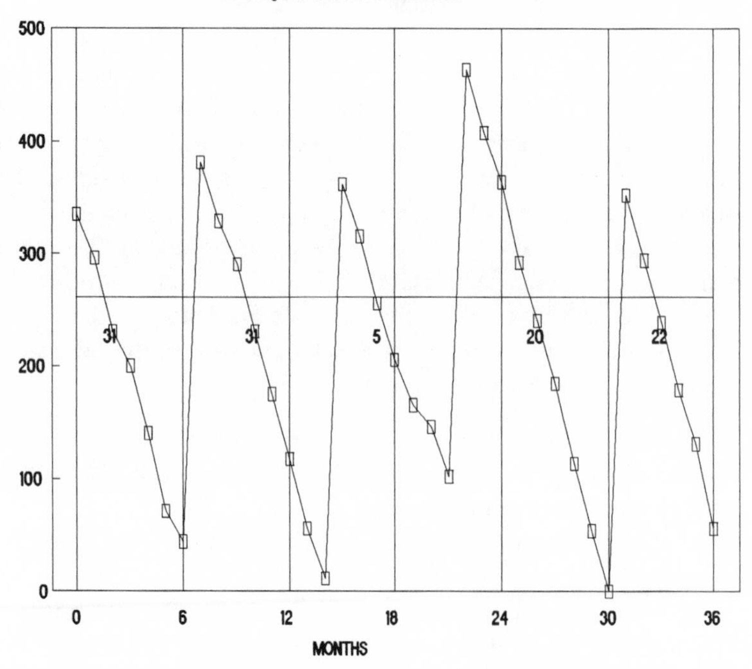

8.1 Overshoot of re-order level demonstrated by LEVELPOL file options VIEWGRAPHS and OVERSHOOT showing that the re-order level is rarely equalled when a replenishment order is raised.

when a replenishment order is raised. It is interesting to note that overshoot of the re-order level is very rarely zero, even though the calculation of the re-order level has so far been based on the fact that stock balances always equal the re-order level exactly at the instant a new replenishment order is raised i.e. that the overshoot is assumed to be zero.

For the situation depicted originally in Fig. 6.3, to establish an estimate of the amount of overshoot that is occurring, if the actual 'vendor service level' is 87.5% (i.e. the probability of stockout per occasion 12.5%) examination of Appendix C provides a corresponding value of the normal variable $u_{0.125} = 1.15$. Substituting this value in Equation [6.1] then provides:

$$M = \bar{D}L + u_{0.125}\,\sigma_d\sqrt{L}$$
$$M = 50 \times 4 + 1.15 \times 15\sqrt{4} = 235 \text{ units}$$

indicating that the effective re-order level is 235 units, 25 units below the theoretical re-order level of 260 units. Hence, in this particular situation the average overshoot would appear to be about 25.

3. Because the re-order level and replenishment quantity are traditionally evaluated using two completely separate methods, there is no recognition of the fact that changes in one parameter could logically be argued to require changes in the other. For instance, if the value of the replenishment quantity were to be doubled (to take advantage for instance of a significant price break) this would halve the number of replenishment orders which would have to be raised and would, therefore, also halve the number of occurrences that the policy was at risk of stockout per annum. Since the policy would then be operating at a much reduced frequency of risk, to maintain a similar level of service this reduced frequency of replenishment could be balanced by lowering the value of the re-order level slightly to increase the risk of stockout on those fewer occasions the policy was at risk.

Because of the lack of relevance (particularly from the customer's point-of-view) of the concept of a service level defined as the probability of not running out of stock on those occasions the policy was at risk of running out, this definition of service has come to be referred to as the vendor service level on the basis that although it might mean something to the stockholder (vendor) who may be aware of the size and frequency of replenishment orders, it certainly means little or nothing to the customers of the stockholding system.

Customer service level defined as the proportion of annual demand met ex-stock per annum

A more meaningful level of service offered by a stock control system would be the proportion of annual demand met ex-stock per annum. If annual demand were known, such a measure of service would indicate to the customer of an inventory system the number of demand units per annum which would not be satisfied ex-stock. Such a measure has proved to be far more meaningful than the vendor service level and has, therefore, been adopted as an alternative measure of service of a stock control system.

To establish theoretically the value of the customer service level it is necessary to establish an estimate of the proportion of annual demand met ex-stock. This is achieved by considering the right-hand tail of the normal distribution of demand during the leadtime existing in the situation where it has now been established that the effective re-order level is only 235 units. This is demonstrated in Fig. 8.2 which shows that although it is

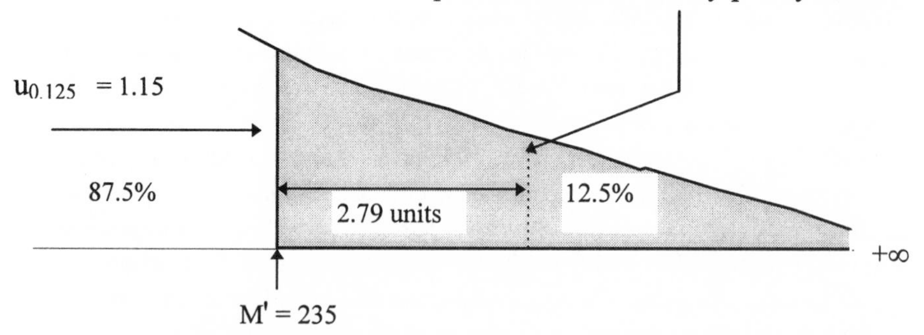

8.2 Distribution of demand during the leadtime showing the right-hand end of the assumed normal distribution between the effective re-order level M′ = 235 and +∞ (plus infinity).

clear that the area (first integral) to the right of the effective re-order level is (100.0 − 87.5) = 12.5% of the total area of the normal probability curve, conceptually in terms of actual units of demand during the leadtime there must also be an average number of units between the value of 235 and plus infinity. This average number of units, if the value could be evaluated, would represent the average number of units of demand during the leadtime in excess of the effective re-order level, i.e. the average number of units of shortage (i.e. demand not met ex-stock) subsequent to the placing of a replenishment order. Given that the number of replenishment orders raised per year is known to be equal to A/q (if A is the annual demand and q the replenishment quantity) in addition to the average shortage per occasion, it would then clearly be possible to establish the number of units per annum which would not be available to customers ex-stock.

To calculate the average shortage per occasion it is necessary to have values of the second integral of the normal probability density function (pdf) between specified limits which would represent the average number of units between those limits. The double (or second) integral of the normal pdf is known as the partial expectation, E(u), and, although not published in many sets of statistical tables, values of the (0,1) normal distribution's partial expectation can be evaluated theoretically as can be seen in Appendix C.

Referring firstly to Fig. 8.2 and then Appendix B, for the situation currently under consideration, a value of the standard normal variable $u_{0.125} = 1.15$, represents the position of the effective re-order level in terms of standard deviation of demand during the leadtime above the average. Again from Appendix C, the value of the partial expectation corresponding to $u_{0.125}$ produces $E(u) = 0.062$ for the (0,1) normal distribution. For the actual distribution of demand during the leadtime with a standard deviation $\sigma_d\sqrt{L} = 30$ units it follows that the equivalent value $E(u)\sigma_d\sqrt{L} = 0.062 \times 30 = 1.86$ units of shortage per occasion a shortage could occur. Given that there are $A/q = 12 \times 50/400 = 1.5$ replenishment orders placed per annum, the annual shortage is given by $1.5 \times 1.67 = 2.79$ units per annum. For an annual demand of $A = 600$ units, a shortage of only 2.79 units means that 597.21 units of demand per annum are met ex-stock thus offering a customer service level defined as the proportion of annual demand met ex-stock of $P' = 597.21/600 = 99.5\%$. This estimated value can be confirmed as the theoretical customer service level of 99.5% shown in Fig. 8.3.

Summarising the above discussion in more general terms, the expected shortage per occasion is given by $E(u)\sigma_d\sqrt{L}$ which for A/q occasions per annum defines the shortage per annum as:

$$\left(\frac{A}{q}\right) E(u)\sigma_d\sqrt{L}.$$

```
C.A.L. Module            OPERATIONS CONTROL - Stock Control                      CDL/96
PARAMETERS DEMAND COSTS SIMULATE  INFORMATION/PRINT VIEWGRAPHS coLOUR QUIT
Specify values of re-order level and replenishment qty

       STOCK CONTROL SIMULATION   :   RE-ORDER LEVEL POLICY  12 years
       =====================================================
   COSTS:                             PARAMETERS:          Actual Theoretical
     Storage(%pa mat'l cost)   22.50%  Re-order level..     235         260
     Material(+lab +o'heads)    1.00   Replen. qty.....     400         400
     Ordering cost/occasion    30.00   Vendor service..   76.5%       87.8%
     Penalty cost/period       30.00   Customer service   98.9%       99.6%
   DEMAND/ MONTH    Specified  Actual SIMULATED   COSTS/ANNUM
     Fcst average       50.00   47.03  Storage costs...........        44.37
     Fcst std dev       15.00   15.80  Ordering costs.........         42.50
     Leadtime (1 to 8)......        4  Penalty costs...........        10.00
   STOCKTURN...............      2.88  Total operating costs....       96.87
   ==========================================================================
   Months  Previous Current Received Current  Backorder O'shoot   Stockout
             stock   demand   orders   stock       qty.   amount     occur.
   ==========================================================================
        0     329      61               268
        1     268      60               208                  27
        2     208      55               153
        3     153      55                98
 File:LEVEL                     Esc toggles menu                           MENU
```

8.3 STOCKMAN's LEVELPOL screen showing that even when the actual re-order level controlling the simulation is set lower than the theoretical value of 260, an acceptable customer service level is still offered.

If the customer service level, P′, is defined as the proportion of annual demand met ex-stock, the defined expected shortage per annum is given by:

$$(1-P')A$$

Equating the two produces:

$$(1-P')A = \left(\frac{A}{q}\right) E(u)\sigma_d\sqrt{L}$$

which allows the customer service level, P′, to be defined by:

$$P' = 1-E(u)\sigma_d\sqrt{\frac{L}{q}} \qquad [8.1]$$

Equation [8.1] not only produces a much more meaningful measure of service but also permits the two controlling parameters of the re-order level policy, the re-order level and the replenishment order quantity to be evaluated jointly.

For instance, if a value of customer service level, P′, is stipulated (on the assumption that the proportion of annual demand met ex-stock is a more meaningful measure of service offered by a stock control system) and the value of q is specified (on the assumption that with a flexible choice of the replenishment order size the stock-holder has a better chance of reducing costs per unit), with the value standard deviation of demand per period σ_d and value of the leadtime, L, specified, it follows that by transposing Equation [8.1] the partial expectation, E(u), can be evaluated using:

$$E(u) = \frac{q(1-P')}{\sigma_d\sqrt{L}} \qquad [8.2]$$

With a value of E(u), the normal variable u can be established from Appendix B and, hence the re-order level M calculated using Equation [6.1]. Thus the values of both parameters controlling the re-order level policy would have been evaluated jointly.

To confirm the validity of Equation [8.1], where it was originally assumed that a re-order level of 260 units offered a vendor service level of 97.7% with the normal variable $u_{0.02275} = 2$ and the partial expectation $E(u) = 0.008$ from Appendix C, then substituting in Equation [8.1] provides:

$$P' = 1 - \frac{0.008 \times 15\sqrt{4}}{400} = 99.9\%$$

which can be confirmed as the theoretical customer service level in Fig. 6.3.

Task 8.1 Establish a costs and demand scenario by specifying:

- $\bar{D}$ an average monthly demand and σ_d a monthly standard deviation of demand;
- L a fixed leadtime expressed in months; and
- storage cost i, ordering cost C_O and material cost C_m.

From this information using the appropriate formulae establish the theoretical re-order level and replenishment quantity to operate a re-order level inventory policy.

Using STOCKMAN's LEVELPOL file, confirm these theoretical results and demonstrate that:

- the actual service levels offered differ from the theoretical; and
- the customer service level is always considerably higher than the vendor service level.

The foregoing discussion, as well as producing a more meaningful measure of service also produces a sensible interrelationship between the ROL and the replenishment quantity. For instance, if the CSL P′ were to be kept fixed in value, with increases in the size of the replenishment quantity the partial expectation E(u) would have to increase. However, examination of Appendix C indicates that an increase in the value of E(u) infers a decrease in the value of u and hence the ROL. Thus, on those fewer occasions (as a result of the increased replenishment quantity) when the inventory system is at risk the system automatically allows the risk of stockout to increase marginally (by lowering the ROL).

Conclusion

Conventional wisdom requires that a measure of service be specified for a stock control system from which the value of the parameters controlling that system can then be established. By specifying a relatively high value of vendor service level, defined as the probability of not running out of stock on those occasions the policy was at risk of running out in practice, because of overshoot of the re-order level a much lower level of service is offered. However, customers' perception of service is measured in terms of the proportion of demand met-ex-stock per annum and it can be shown that this measure of customer service level is always much higher than the value of vendor service level. Thus, although the stockholder (or vendor) offers customers nowhere near as high a service level as specified, in terms that customers understand they actually receive a higher level of service. Perhaps this is one of those rare circumstances where several wrongs make a right!

The foregoing argument suggests that it would be logical to set the re-order level lower than that suggested theoretically using vendor service level criteria and Fig. 8.3 demonstrates this contention by illustrating that in the demand and cost situation discussed in this chapter a re-order level set at 235 on the basis of offering only a theoretical vendor service level of 87.8%, and achieving an actual 76.5%, offers a very acceptable customer service level of 98.9%.

File from OPSCON's package STOCKMAN associated with this chapter

LEVELPOL file – see page 90 for details.

Part III

Inventory control – the re-order cycle inventory policy

9

Establishing the value of the review period and maximum stock level

Introduction

The re-order cycle inventory policy (sometimes known as the period review system) is the next most commonly used inventory control policy after the re-order level policy. It operates on the basis that replenishment orders of a variable size are placed at regular intervals of time, R. When a replenishment order is raised, it is evaluated as that size of order which would bring the stock balances up to a maximum level, S, if replenishment were instantaneous (i.e. leadtime was zero). This ensures that when stock-on-hand (i.e. physical stock held plus outstanding replenishments less committed stock) is low at review relatively large replenishment orders are placed and conversely when stock-on-hand is high at review relatively small replenishment orders are placed. This effect helps to keep stock balances stable.

Because in this policy replenishment orders are placed regularly, thus allowing for multiple item orders on suppliers (i.e. one order for possibly many items), it is generally assumed that the cost of raising individual replenishment orders on suppliers within the re-order cycle policy is cheaper than when operating a re-order level policy.

Selecting the review period in the re-order cycle inventory policy

The re-order cycle policy is controlled entirely by the review period, R, and the maximum stock level, S, in that on each occasion the stock-on-hand is reviewed a replenishment order equal to the maximum stock level, S, less the current level of stock is placed. Because the review period is a measure of time, it is often chosen on an arbitrary basis to fit in with a wider time framework and therefore could be chosen to be a calendar month (or planning period), a quarter (three months) or even a year.

However, it is possible to use an approach similar to that used in Chapter 7 to formulate the Economic Order Quantity and within this re-order cycle policy establish m_o, the 'economic number of reviews per annum' and from this deduce the 'economic review period', R_o, which would minimise the annual operating cost.

Developing a similar approach (see page 95) for formulating the annual operating cost, C, in terms of the cost of ordering replenishment orders and the cost of storage per annum for the re-order cycle policy produces:

$$C = mC_o + \left(\frac{A}{2m} + S\right)iC_m \qquad [9.1]$$

where:

m is the number of reviews made per year,
C_o is the cost of raising a replenishment order (say 30),
A is the annual demand (say 600),
S is a fixed safety stock, and
iC_m is the annual unit storage cost (say at 22.5% pa 0.225*1 = 0.225).

Then, by differentiating the annual operating cost, C, (as defined by Equation 9.1) with respect to the number of reviews, m, defines dC/dm as:

$$\frac{dC}{dm} = C_o - \left(\frac{AiC_m}{2m^2}\right) \qquad [9.2]$$

Hence, the optimal frequency, m_o, of placing replenishment orders which minimises the annual operating cost, is then defined by dC/dm = 0 from which the 'economic number of reviews per annum' m_o is defined as:

$$m_o = \sqrt{\frac{AiC_m}{2C_o}} \qquad [9.3]$$

Where the period of time (or time unit) on which the demand average and standard deviation are based and also the leadtime is measured is calendar months, the review period, R_o, which minimises the annual operating cost based on a twelve month year is given by:

$$R_o = \frac{12}{m_o} = \frac{12}{\sqrt{\frac{AiC_m}{2C_o}}} \qquad [9.4]$$

which, for the situation under consideration produces:

$$R_o = \frac{12}{m_o} = \frac{12}{\sqrt{\dfrac{600*0.225*1}{2*30}}}$$

$$R_o = 8 \text{ months}$$

The value of R_o developed using this approach should only be regarded as an indication as to an appropriate review period. In practice the choice of a review period for a re-order cycle policy will need to align with other overall time factors.

Setting the maximum stock level

Examining the stock balances for the re-order cycle policy as shown in Fig. 9.1, it should be noted that the decision to place a replenishment order (indicated by ord) at position A (period 24) effects the outcome at position C (period 36) with regard to the possibility of stockout occurring prior to the receipt of the replenishment order (indicated by rec) rather than at position B (period 28). This is because while the outcome at C, in terms of a potential stockout is a function of the stock position at B , the stock position at B is a function of the decision as to whether to place a large or small replenishment order at the previous review which occurred at period A. Since the periods A and C are separated by (R + L) time periods, this time duration (R + L) can be interpreted as the period of risk for the re-order cycle policy.

If it is accepted that (R + L) be regarded as the period of risk for a re-order cycle policy and if the maximum stock level, S, can be regarded as equivalent to a re-order level which stock balances are *always* below, then Equation [6.1] (see page 86) – originally proposed for establishing the value of the re-order level within a re-order level policy – can now be modified to establish the value of the maximum stock level, S, within a re-order cycle policy by replacing M with S and L with (R + L) such that:

$$S = \bar{D}(R + L) + u_{0.02275}\,\sigma_d\sqrt{R+L} \qquad [9.5]$$

For the demand situation dealt with previously with $\bar{D} = 50$, $\sigma_d = 15$, $L = 4$ and with now in addition the review period $R = 8$, for a 97.7% 'vendor service level' – probability of not running out of stock on those occasions the policy is at risk of running out – for which $u_{0.02275} = 2$ from Appendix C, the maximum stock level S developed from Equation [9.5] would produce a value of 704 units as shown below:

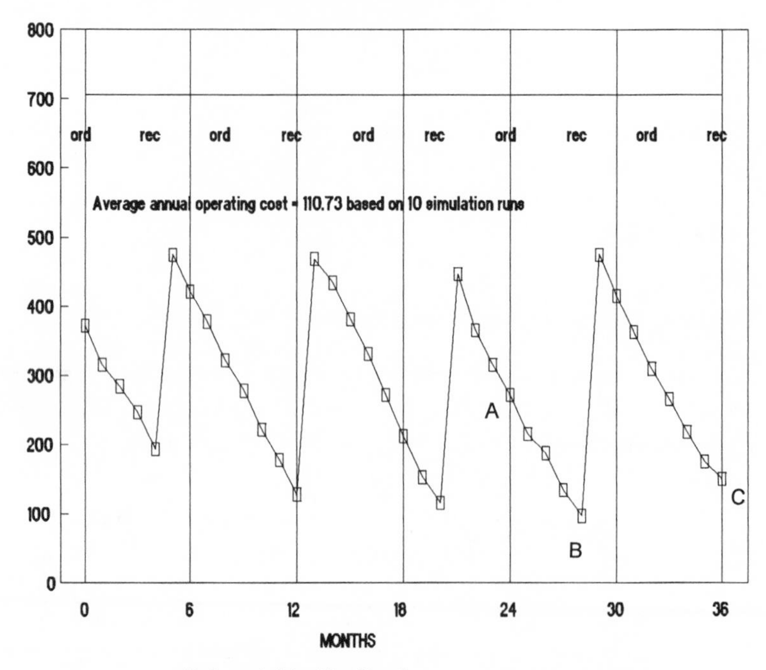

9.1 Re-order cycle policy stock balances for initial three year period.

$$S = 50 \times (8 + 4) + 2 \times 15 \times \sqrt{8+4} = 600 + 104$$
$$S = 704 \text{ units}$$

The value of S = 704 units is the actual level used in producing the stock balances shown in Fig. 9.2 (generated using the TOTAL option within VIEWGRAPHS). This shows a less detailed view of the same situation shown in Fig. 9.1 but taken over a longer (twelve year) period. It also confirms that a maximum stock level of 704 units appears to offer a reasonable level of service since no stockouts occur in that twelve year period.

Readers should also note in Fig. 9.3, which is STOCKMAN's CYCLEPOL file's screen display, that the actual vendor service level is not significantly lower than the theoretical level of 97.7% – as was invariably the case in the re-order level policy – and is indeed 100% in this particular case, since no stockouts occur in the twelve years of the simulation period.

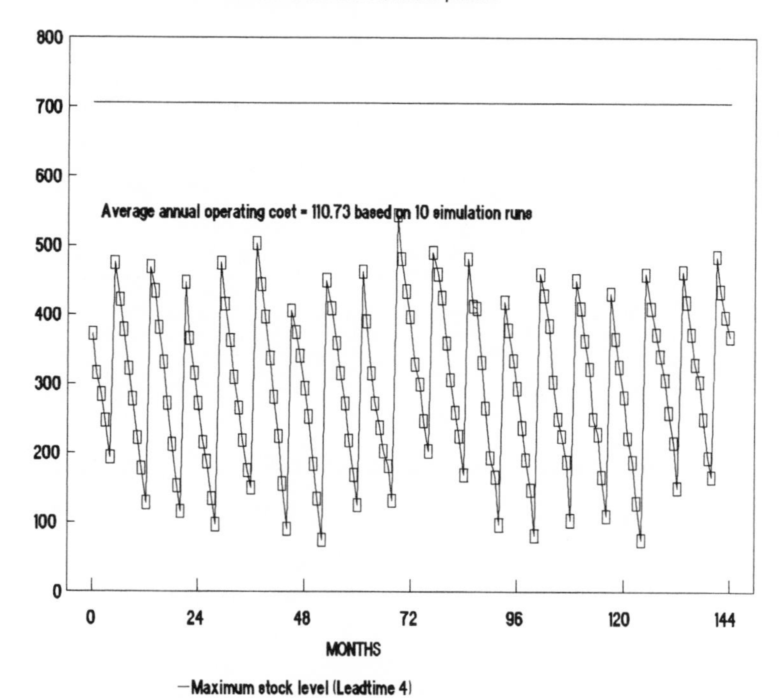

9.2 Re-order cycle policy stock balances for initial twelve year period.

Task 9.1 Using STOCKMAN's CYCLEPOL file examine the practical range of maximum stock level values for the current demand, review period and leadtime using the PARAMETER option followed by the MAXIMUM option. What assumptions has the package made in determining the maximum and minimum values of the range?

Within the practical range of the maximum stock level:

1. examine the operating costs and establish where costs are minimised and comment on the reasons why operating costs increase at either end of the range;
2. record the values of the theoretical and actual vendor service levels and comment on how these vary with the maximum stock level settings.

```
C.A.L. Module          OPERATIONS CONTROL - Stock Control                    CDL/96
PARAMETERS DEMAND COSTS  SIMULATE  INFORMATION/PRINT VIEWGRAPHS coLOUR QUIT
Specify values of maximum stock level or review period

      STOCK CONTROL SIMULATION   :   RE-ORDER CYCLE POLICY  12 years
           ===================================================
COSTS:                               PARAMETERS:          Actual Theoretical
  Storage(%pa mat'l cost)     22.50% Max. stock level      704         704
  Material(+lab +o'heads)      1.00  Ave. repl. qty..      378          NA
  Ordering cost/occasion      30.00  Vendor service..   100.0%       97.7%
  Penalty cost/period         30.00  Customer service   100.0%       99.7%
DEMAND/ MONTH      Specified  Actual SIMULATION  COSTS/ANNUM
  Average.......      50.00   50.25  Storage costs...........       62.93
  Std. deviation      15.00   15.69  Ordering costs.........        45.00
  Review|Leadtime         8       4  Penalty costs..........
STOCKTURN................      2.17  Total costs............       107.93
===========================================================================
Months  Previous  Current Received  Current Backorder Orders Periods of
           stock   demand   orders    stock      qty.  placed  stockout
===========================================================================
     0      422       55               367              337
     1      367       74               293
     2      293       65               228
     3      228       76               152
File:CYCLEPOL                   Esc toggles menu                       MENU
```

9.3 STOCKMAN's CYCLEPOL screen display showing information relating to demand situation where average demand $\bar{D} = 50$ and standard deviation of demand $\sigma_d = 15$ with a review period of 8 and a leadtime of 4. The actual maximum stock level controlling the simulation is set equal to the theoretical value of 704 computed using Equation [9.5].

Establishing the customer service level defined as the proportion of annual demand met ex-stock per annum within a re-order cycle policy

Using similar arguments as established previously for the re-order level policy (see page 114) in the case of the re-order cycle policy situation being considered here, the period of risk for this policy is $(R + L)$ and the policy is at risk of a stockout subsequent to $m = 12/R$ replenishment orders being placed per annum. Given that the expected shortage per occasion is $E(u)\sigma_d \sqrt{R+L}$, the expected shortage per annum would be given by:

$$\left(\frac{12}{R}\right) E(u)\sigma_d \sqrt{R+L}$$

If again the customer service level, P', is defined as the proportion of annual demand met ex-stock per annum the defined expected shortage per annum is given by:

$$(1 - P')A$$

then equating the two produces:

$$(1 - P')A = \left(\frac{12}{R}\right) E(u)\sigma_d \sqrt{R+L}$$

given that $A = 12\bar{D}$, then P′ can be defined for the re-order cycle policy as:

$$P' = 1 - \frac{E(u)\sigma_d \sqrt{R+L}}{R\bar{D}} \quad [9.6]$$

With the appropriate value of E(u) for $u_{0.02275}$ of 0.008 from Appendix C (there being no overshoot effect to consider in the re-order cycle policy) for the situation shown in Fig. 9.3 P′ is evaluated as:

$$P' = 1 - \frac{0.008 \times 15 \times \sqrt{12}}{8 \times 50} = 0.9989$$

offering a theoretical customer service level of 99.89% (shown as 99.7% in Fig. 9.3 due to the small error caused in generating the value of E(u) using a fourth order polynomial approximation). Again (as in the re-order level policy) the customer service level is generally higher than the vendor service level although much closer in value within this policy.

Linking the forecast of demand to the setting of the re-order level

Within the re-order level policy simulation file CYCLEPOL it is possible to change the average and standard deviation of demand to any set of chosen values. However, in a practical inventory control environment, these values would be established as a result of an earlier analysis of some real demand data with a forecasting package. To mimic this facility within OPSCON, when quitting any of FOREMAN's forecasting files the user is offered the option of exporting the results of the forecasting analysis to a temporary file for subsequent importation to either of the inventory control files CYCLEPOL or LEVELPOL (see Chapter 7). When quitting in particular from FOREMAN files whose forecasting models assume growth or seasonality, because the forecast ahead varies over the forecast horizon, the user is asked to specify from which period ahead the forecast value is to be taken.

Practitioners with a real stocked item to consider should recognise that when the forecast data is imported into inventory policy simulation files this does not include any cost data. To establish a realistic cost environment it will at least be necessary to change the material costs via the COSTS and MATERIAL options; the storage and ordering costs default values being a realistic 22.5 and 30% respectively.

Task 9.2 Initially use any of FOREMAN's forecasting files to analyse either demand data provided by the appropriate DEMONSTRATION option or demand data previously saved as an ENTIRE file using the DATAHELP facility. When using the QUIT option to terminate forecasting analysis, use the subsequent EXPORT option to retain the results of the analysis before proceeding automatically to STOCKMAN.

Subsequently, load STOCKMAN's CYCLEPOL file and use the DEMAND and FORECAST options to load the retained forecast data which effectively becomes the average demand ($\bar{D}$) and the standard deviation of demand (σ_d). Confirm that the maximum stock level value evaluated is appropriate for the original demand data and justify why.

Conclusion

The setting of the review period, R, within a re-order cycle policy will usually be on the basis of choosing a time period which fits in with an overall time framework. However, by adopting an approach similar to the Economic Order Quantity approach discussed in Chapter 6, it is possible to show that an economic review period can be established. However such a review period should only be considered as indicative.

Having specified the review period for the re-order cycle policy, the remaining parameter which controls the operation of the policy is the maximum stock level, S, which is used to evaluate the size of replenishment orders at review. The value of the maximum stock level can be established using a vendor service level approach and the resulting theoretical and actual values are likely to be much closer than was the case for the re-order level policy. Subsequent considerations show that the customer service level will be higher than the vendor service level both in theory and in practice.

File from OPSCON's package STOCKMAN associated with this chapter

CYCLEPOL file – a complex, multi-level menu driven file which simulates the stock balance of the re-order cycle policy.

The controlling parameter of the policy, namely the maximum stock level, is evaluated in the light of demand and costs conditions together with a specified review period and the user can either use the recommended values or specify alternative values.

The demand environment can be specified in terms of the average and standard deviation of demand per unit time and the duration of the replenishment leadtime. Alternatively the demand environment can be established as a result of a previous forecasting analysis from which the relevant forecasting information can be imported.

The file contains a comprehensive cost model and users can examine the results of the simulation model through a variety of graphs which display:

- initial and overall stock balances together with detail of stockouts where these occur;
- the demand environment – in terms of a scatter plot of demand orders and a frequency distribution of those orders; and
- a breakdown of storage, ordering, penalty and purchase/acquisition costs.

All simulations can be run under the assumptions that backordering (the assumption that demand orders that cannot be met immediately are nevertheless accepted thus inferring negative stock) can either be allowed or prohibited. To provide relatively stable cost analyses, when a simulation is run the average costs resulting from ten simulation runs are reported.

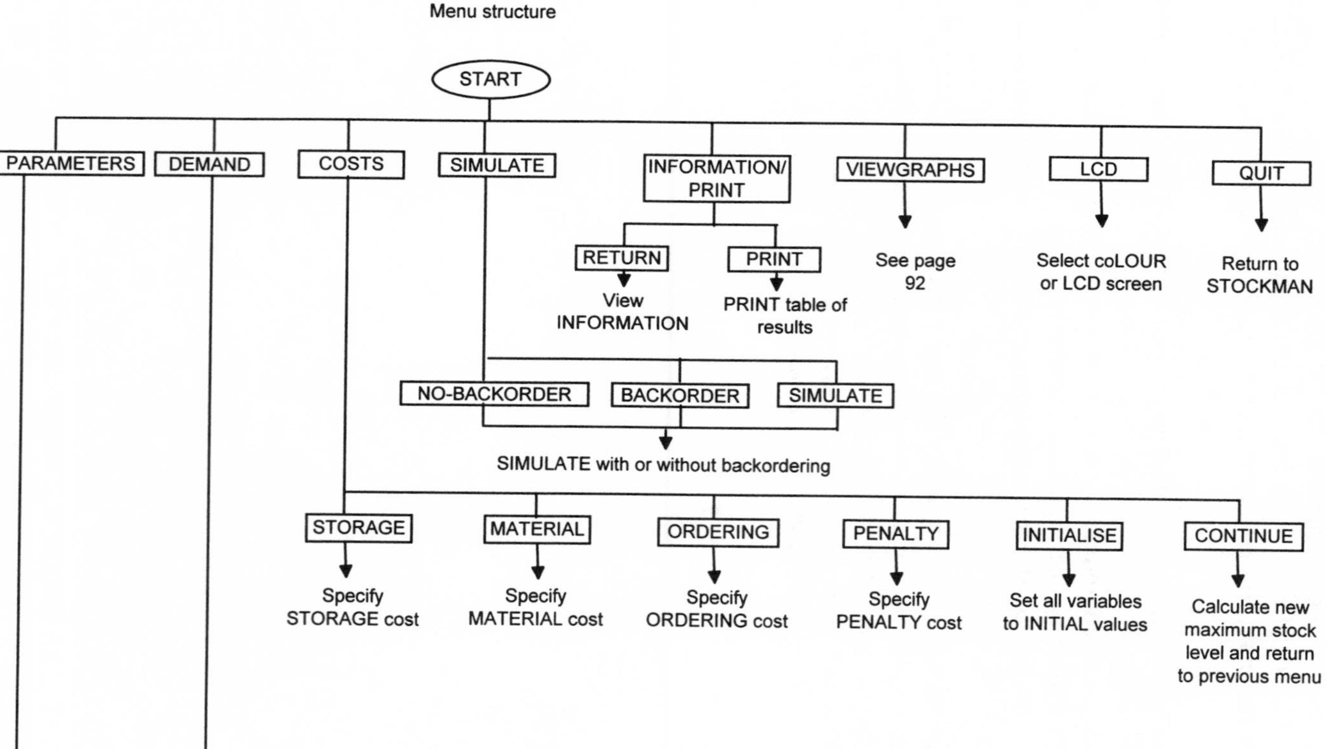
CYCLEPOL
Menu structure
START
PARAMETERS
DEMAND
COSTS
SIMULATE
INFORMATION/
PRINT
VIEWGRAPHS
LCD
QUIT
RETURN
View
INFORMATION
PRINT
PRINT table of
results
See page
92
Select coLOUR
or LCD screen
Return to
STOCKMAN
NO-BACKORDER
BACKORDER
SIMULATE
SIMULATE with or without backordering
STORAGE
Specify
STORAGE cost
MATERIAL
Specify
MATERIAL cost
ORDERING
Specify
ORDERING cost
PENALTY
Specify
PENALTY cost
INITIALISE
Set all variables
to INITIAL values
CONTINUE
Calculate new
maximum stock
level and return
to previous menu

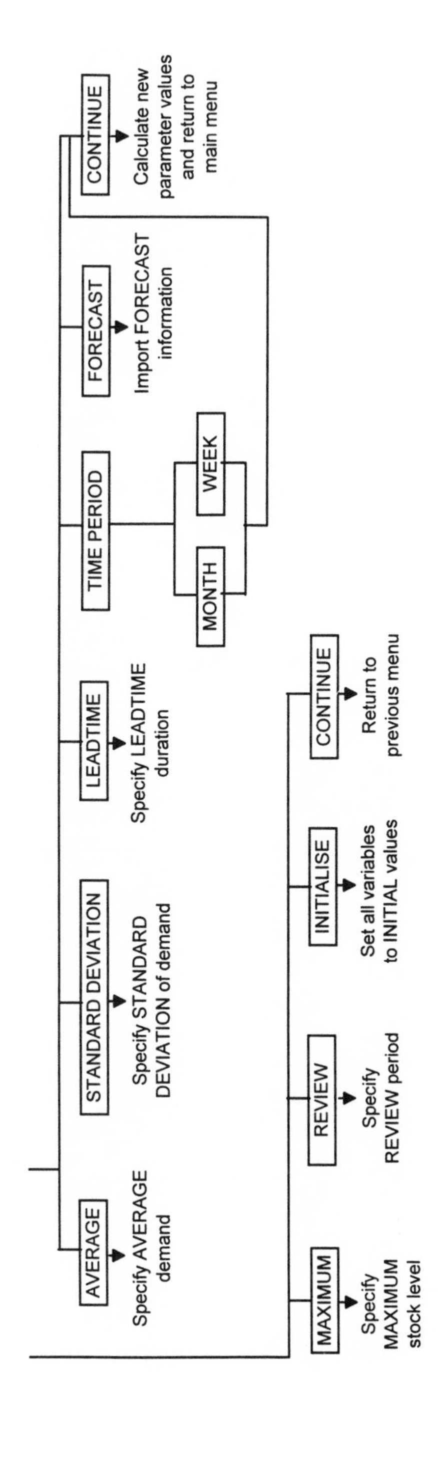
AVERAGE
Specify AVERAGE demand
STANDARD DEVIATION
Specify STANDARD DEVIATION of demand
LEADTIME
Specify LEADTIME duration
TIME PERIOD
MONTH
WEEK
FORECAST
Import FORECAST information
CONTINUE
Calculate new parameter values and return to main menu
MAXIMUM
Specify MAXIMUM stock level
REVIEW
Specify REVIEW period
INITIALISE
Set all variables to INITIAL values
CONTINUE
Return to previous menu

Part IV

Conclusion

10

Selecting the most appropriate inventory control policy

Introduction

Within any inventory control system involving many stocked items, it is clearly not sensible to treat all items as being equally important in terms of selection of an appropriate inventory control policy; any more than for selecting a group of forecasting models – as was explored in detail in Chapter 5. In deciding which type of inventory control policy might be appropriate for each stocked item, in broad terms there are three possibilities and hence the Pareto or ABC categorisation scheme will again be considered as an appropriate allocation vehicle. Readers who have not covered the material related to this categorisation scheme should refer to Chapter 5, pp 69–72.

With regard to the inventory control policies which are available for allocation to appropriate stocked items, the choice lies between:

1. The re-order level policy (or fixed order quantity system) within which the re-order level and replenishment quantity are established from information derived from the value of the duration of the replenishment leadtime together with forecast and cost information.
2. The re-order cycle policy (or periodic review system) within which the maximum stock level (which is used as a basis for establishing the size of replenishment orders) is established from information derived from the duration of the replenishment leadtime and the policy's review period together with forecast information.
3. The two-bin policy i.e. a re-order level policy operated on the basis that when the first bin is emptied a replenishment order is placed and further demand met from a second bin whose size effectively becomes a re-order level. This inventory policy does not require a formalised forecasting procedure but does assume that the quantity of stock held is a function of the volume of space occupied and clearly this is only likely to be true for relatively small, homogeneous items.

Table 10.1 Cost characteristics of re-order level and re-order cycle inventory policies

Common cost features in both policies	Re-order level policy	Re-order cycle policy
Storage costs pa	Relatively low due to shorter period of risk, L	Relatively high due to longer period of risk (R + L)
Ordering costs pa	Relatively high due to irregular raising of replenishment orders	Relatively low due to regular raising of replenishment orders and the opportunity for multi-item orders
Penalty costs pa	Similar penalty costs for both policies when offering a similar level of customer service	
Operating costs pa (Storage + Ordering + Penalty)	Similar for both policies when offering similar levels of customer service since the higher ordering costs and lower storage costs of the re-order level policy tend to be balanced by the higher storage costs and lower ordering costs of the re-order cycle policy.	
Purchase/acquisition costs pa	Identical purchase costs for both policies given the purchase price per unit is unaffected by the different ordering regimes i.e. no discount for increased replenishment order sizes within the re-order level policy or no discount for the regular placing of orders within the re-order cycle policy.	
Total costs pa (Operating + Purchase)	For similar levels of customer service, total costs pa for both policies not significantly different if control parameters correctly adjusted	

Selection of inventory control policy based on type of stocked item – Pareto analysis

The allocation of appropriate inventory policies to 'A', 'B' and 'C' stocked items is not as clear cut as was the allocation of forecasting models discussed in Chapter 5. Both policies have advantages and disadvantages and Table 10.1 summarises some of the essential differences between the re-order level and re-order cycle policy and shows that both policies tend to operate at similar annual costs when offering similar levels of customer service.

Allocation of inventory control policy(ies) to 'A' and 'B' items

Because there does not appear to be a definite advantage in allocating either the re-order level or re-order cycle policy to 'A' or 'B' items, the general recommendation would be that since both 'A' and 'B' items are clearly of significant importance (relative to 'C' items) they should be controlled by a formalised inventory control policy fed with information derived from a formalised forecasting procedure.

In attempting to be more specific as to which of the two major inventory control policies is more appropriate, Waters[11] has indicated that:

> The main benefit of a periodic review system (i.e. re-order cycle policy) is that it is simple and convenient to administer. There is a routine where stock is checked at regular times, orders are placed, delivery is arranged, goods arriving are checked, and so on. This is particularly useful for cheap items with high demand. The routine also means that the stock level is only checked at specific intervals and does not have to be monitored continuously.

On the basis that 'B' items are cheaper than 'A' items, Waters' argument would suggest on balance that 'B' items should be controlled by the re-order cycle policy.

With regard to the re-order level policy, Waters argues that:

> A major advantage of the fixed order quantity systems is that orders of a constant size are easier to regulate than variable ones. Suppliers know how much to send and the administration and transportation can be tailored to specific needs (perhaps supplying a truck load at a time). Also a fixed order system has more flexibility to suit order frequency to demand.

Such advantages might indicate that the re-order level policy is marginally more suitable for 'A' items. This assertion could be said to have added emphasis now that one of the basic criticisms of the re-order level policy of not being able to identify items which were rapidly becoming obsolete has been overcome by computerisation. This criticism used to be levelled at the re-order level policy on the basis that, with little or no demand, re-order levels were not broken and therefore no message emanated from the system indicating that demand for the item had nearly ceased. All computerised inventory control systems are now capable of producing exception reports for slow movers such that this inability to identify potentially obsolete items simply because re-order levels are not being broken should no longer be a significant problem associated specifically with this policy.

Allocation of inventory control policy to 'C' items

In considering the most appropriate inventory control approach which could be associated with 'C' items, the fact that 'C' items represent the majority of items in stock control systems but a very small proportion of overall value must raise the issue of whether a formalised inventory control system is appropriate for these items at all. As was discussed in Chapter 5 with regard to the allocation of appropriate forecasting models to 'C' items, because these items represent such a small proportion of the value invested in stocked items but a large proportion of the actual number it may be sensible in practice to control these items simply by overstocking. This repeats the earlier argument that in approximate terms 20% overstocking of 5% of the investment in stocked parts only incurs an approximate cost of 1% overall. If the cost of both a formalised forecasting system and the cost of the subsequent linking of information derived from the forecasts to establish the correct parameters of either a re-order level or re-order cycle inventory policy is greater than the 1% of stock investment – it could well be that an alternative approach should be considered.

Such an approach could be the so called two bin policy which is effectively a re-order level policy within which:

- parts are held in two bins (or one bin with a marked level);
- it is assumed that parts are relatively small and homogeneous and therefore the number of parts held are a function of the volume occupied;
- there is no formalised forecasting system; and
- replenishment orders are raised when the first bin is emptied (or parts fall below the marked level in a single bin) both being regarded as being equivalent to the re-order level.

Hybrid or specialised inventory policies

Although the re-order level and re-order cycle inventory policies are the basis of virtually all formalised stock control systems, there are clearly situations where such fundamentally simple systems can on the one hand either be improved or on the other are clearly inappropriate. In this concluding chapter an example of a hybrid policy know as the (s,S) policy is considered together with an inventory policy designed specifically for slow moving items for which neither the re-order level nor re-order cycle policies are appropriate.

The (s,S) policy – a hybrid inventory policy based on both the re-order level and re-order cycle policies

A criticism of the re-order cycle inventory policy could be that since replenishment orders are placed at each and every review, should the current stock level at review be high, a relatively small replenishment order will be placed. One method of preventing this happening would be to introduce a third control parameter (usually referred to as 's') into the system with a similar function to a re-order level such that the rule for raising a replenishment order becomes 'Place a replenishment order at review only if the current stock level is below s'.

Proponents of such an inventory policy – which by defining the maximum stock level as S has become know as an (s,S) policy – have suggested that it could be made to operate as a re-order cycle policy by raising s to a high enough level that current stocks were always below s at review. Similarly by lowering the value of s, this (s,S) policy could be made to operate more like a re-order level policy since the raising of replenishment orders would be more dependent on the value of s than the review period. Given that the (s,S) policy can be operated like either a re-order level or re-order cycle policy simply by modifying the value of s, this suggests that it should be possible at some position between these extremes to operate the (s,S) policy more effectively than either of the two policies from which it has been derived. Several theoretical studies of the (s,S) policy under specific conditions have supported this contention but it is generally agreed that with three controlling parameters, namely:

1. the equivalent of the re-order level s;
2. the maximum stock level S; and
3. the review period R;

achieving the correct combination of all three to optimise the operation of the (s,S) policy is a relatively complex task. For this reason the STOCKMAN element of the OPSCON package does not include this policy. However, Fig. 10.1 demonstrates how such a policy would operate in a situation where with a value of s = 315 and S = 704 the (s,S) policy is seen to be operating most of the time as a re-order cycle policy with only an occasional review not triggering the placing of a replenishment order due to a relatively high stock level at review. This can be seen to occur at the review highlighted at a time identified as A where the non-placement of a replenishment order has actually caused a subsequent stockout. This illustrates the point made previously concerning the difficulty of obtaining the correct combination of the controlling parameters to operate this policy optimally.

Inventory policy for slow moving items such as engineering spares

When the demand for an item is defined in terms of its interval between issues and that interval could be of the order of twenty years, clearly one is no longer talking about fast moving items to which most of this book has been dedicated. In such situations it is not so much a matter of establishing an appropriate re-order level, replenishment quantity or maximum stock level but rather whether even one unit of such an item

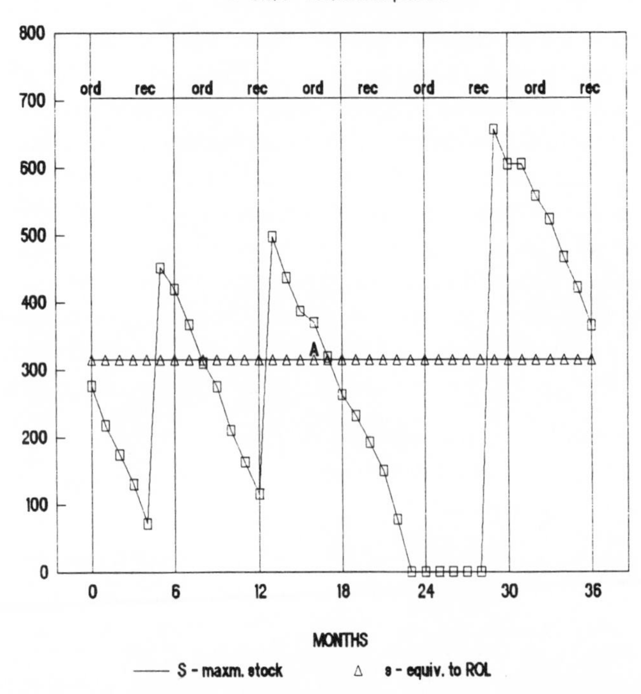

10.1 Stock balances for the hybrid (s,S) inventory policy within which replenishment orders are raised at review only if the stock level is below s and indicating that an order is not raised at time A.

should be held in stock. Such slow moving items, of which engineering spares are a good example, are exemplified by:

- a demand specified by the average interval between issues of a single unit of stock where the average interval could be between six months and twenty years;
- a potentially high storage cost since the few units held in stock might well be held for many years;
- a high penalty cost since the stocked item might well be a spare part whose non-availability could involve a loss of earnings from an expensive piece of capital equipment or a high cost of providing alternative services; and
- a low annual ordering cost since the spreading of a cost of about £30 over several years pales into insignificance compared with the storage and penalty cost mentioned earlier.

The model proposed here for coping with slow moving stock items is based on the work of Karush[4] and assumes the following notation:

γ average interval between demands
C_o average cost of raising a replenishment order
C_m material cost or purchase price
C_s stockout cost (the average extra cost each time a demand cannot be met)
i annual storage cost expressed as a percentage of material cost or purchase price
L average procurement leadtime (delay between raising an order and its receipt)
P(n) probability that there will be just n demands during the leadtime

It then follows that if it is assumed that the demand for the slow moving part or spare occurs at random during the leadtime and the distribution of that demand follows a Poisson distribution, then:

$$P(n) = \frac{(L/\tau)^n e^{-L/\tau}}{n!} \qquad [10.1]$$

For a re-order level inventory policy with a unit order quantity (i.e. only one unit of the stocked item is ordered whenever a replenishment order is raised) Karush[4] has shown that the average annual cost C_N in the steady state with a maximum number of units of stock held defined as N is given by a combination of annual storage, penalty and ordering cost:

$$C_N = iC_p\left[N - \frac{L}{\tau}\left\{\sum_{n=0}^{N-1} P(n) / \sum_{n=0}^{N} P(n)\right\}\right] + \frac{C_s}{\tau}\left\{P(N) / \sum_{n=0}^{N} P(n)\right\}$$
$$+ \frac{C_{os}}{\tau}\left\{1 - P(N) / \sum_{n=0}^{N} P(n)\right\} \qquad [10.2]$$

This apparently complex cost formulation as subsequently adapted by Mitchell[7] has been incorporated in STOCKMAN's file SPARE whose display screen is shown as Fig. 10.2 . This displays the required input variables to the left-hand side of the screen and the annual storage, penalty and ordering costs for a limited range of stocked units from zero to four to the right. The number of stocked units which produces the minimum total annual cost is also highlighted. A graph of the total costs for this limited range of stocked units can be generated by the VIEWGRAPH options and is shown as Fig. 10.3.

```
C.A.L. Module          OPERATIONS CONTROL - Stock Control                 CDL/96
INTERVAL LEADTIME COSTS VIEWGRAPH PRINT GRAPHSAVE colOUR QUIT
Specify average interval between issues (years)

              *** SLOW MOVING SPARES - STOCK HOLDING POLICY ***
              ==================================================
                                              ANNUAL COST OF POLICY
  Average interval               |           ---------------------------
  between issues                 |   Spares Storage Penalty Ordering Total
  in (years)........        20   |    Held    Cost    Cost    Cost    Cost
  Average cost of                |   ======================================
  raising an                     |      0       0     2500       0    2500
  order/occasion.....       30   |=>    1     352      156       1     509 <=
  Penalty cost per               |      2     748        5       1     755
  stockout..........     50000   |      3    1125        0       1    1127
  Purchase price of              |      4    1500        0       1    1502
  spare part.........     1500   |
  Annual holding int-            |
  rest rate (%)......    25.0%   |
  Average delivery               |
  leadtime (months)..       16   |
  Ratio, penalty cost            |
  to purchase price..     33.3   |

 File:SPARE                      Esc toggles menu
```

10.2 STOCKMAN's SPARE file's screen display showing storage, penalty and ordering costs for a slow moving stocked item.

Conclusion

This chapter has discussed the suitability of different inventory control systems for different types of stocked items. Although it is not clear as to which of the re-order level or re-order cycle policies is more appropriate for either 'A' or 'B' items, the relative importance of these items suggests that they should be controlled by a formalised inventory control system linked to a formalised forecasting system. The concept of a hybrid inventory control system known as the (s,S) policy which can be made to operate as either of the two policies discussed previously and feasibly could be operated at a marginally cheaper cost while offering the same level of customer service is also considered.

For 'C' items it is argued that on a cost benefit approach it might be sensible to stock the many but relatively unimportant items using a two bin inventory control system and not employ a formalised forecasting system at all.

This chapter has also discussed the fact that neither the re-order level nor re-order cycle inventory policies are appropriate for controlling slow moving stock items such as engineering spares. Because for this category of item the demand is more likely to be Poisson rather than normal, and also because the penalty cost has an increased significance compared with

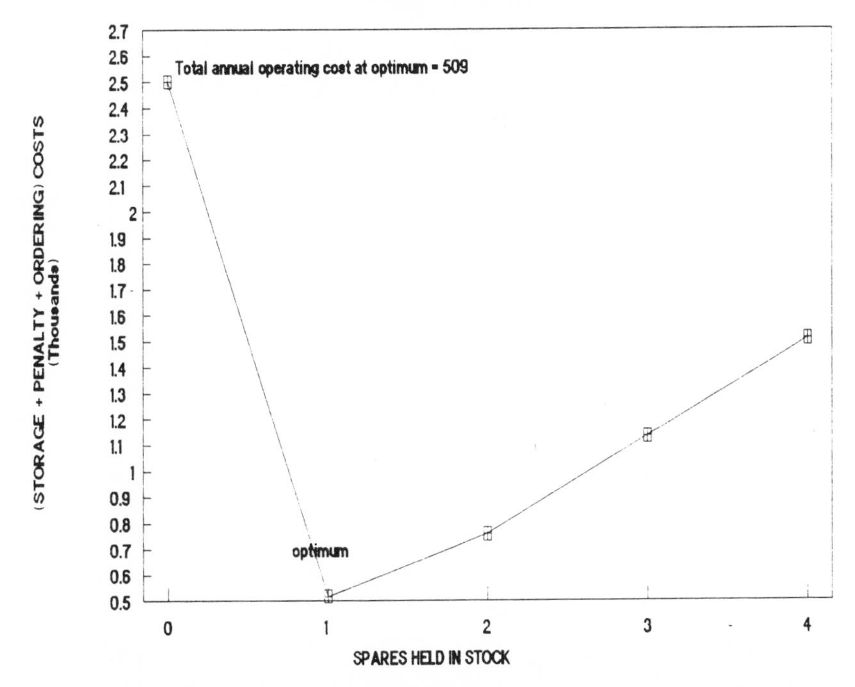

10.3 Total annual operating costs for a slow moving stocked item as generated by STOCKMAN's SPARE file.

the ordering cost, an alternative inventory control policy which indicates the number of units of stock which should be held to minimise costs is proposed.

Files from OPSCON's package STOCKMAN associated with this chapter

PARETO file – see page 76 for details.

SPARE file – a simple, single-level menu driven file which contains a cost model based on the assumption that demand for slow moving spares is distributed as a Poisson distribution.

Options available to the user are:

- specification of the average interval between issues which can be as long as twenty years;
- specification of the supplier's replenishment leadtime in months;

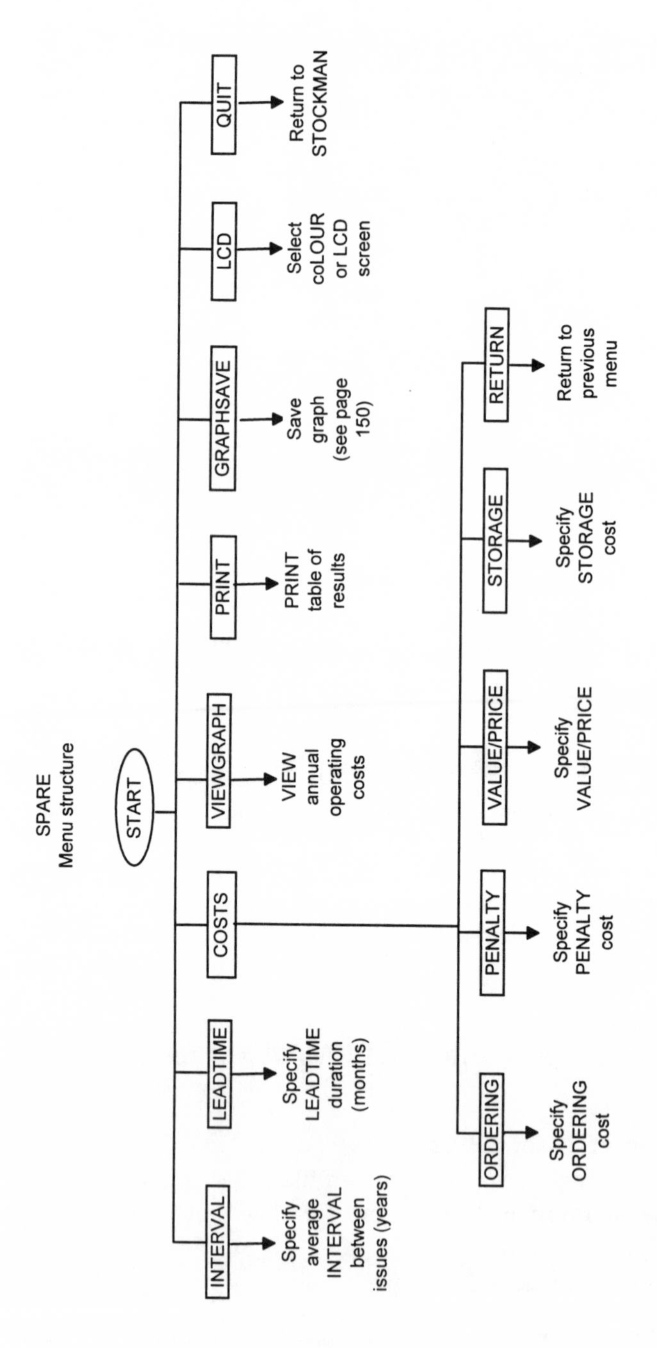
SPARE
Menu structure
START
INTERVAL
Specify average INTERVAL between issues (years)
LEADTIME
Specify LEADTIME duration (months)
COSTS
VIEWGRAPH
VIEW annual operating costs
PRINT
PRINT table of results
GRAPHSAVE
Save graph (see page 150)
LCD
Select coLOUR or LCD screen
QUIT
Return to STOCKMAN
ORDERING
Specify ORDERING cost
PENALTY
Specify PENALTY cost
VALUE/PRICE
Specify VALUE/PRICE
STORAGE
Specify STORAGE cost
RETURN
Return to previous menu

- specification of ordering, penalty and storage costs together with the price or value of the spare.

Supplied with the information described above, the file recommends how many spares should be stocked over a limited range from 0 (zero) to 4 (four) and displays annual operating costs over this range so that the sensitivity of these costs can be assessed.

Appendix A: The OPSCON package

Installing and running OPSCON within a DOS environment

Although the OPSCON package can be run under DOS on a PC using the floppy disk supplied with this book by invoking the command:

A:\OPSCON

the package will run many times faster if loaded on to the user's hard disk.

To install OPSCON on a hard disk (generally specified as drive C:\) in a directory C:\OPSCON; initially insert the disk supplied in the floppy disk drive and from the DOS prompt type the command:

A:\INSTALL

This command will copy all the necessary files to a directory C:\OPSCON on the hard disk and will also automatically start the program such that the OPSCON opening screen display should appear as in Fig. A1.

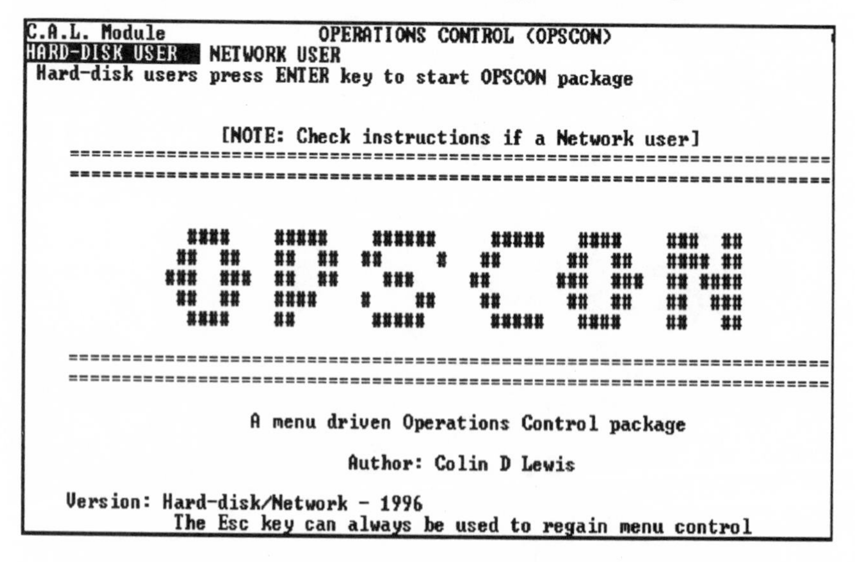

A1 OPSCON's opening screen at which the user specifies whether a HARD DISK or NETWORK user.

The OPSCON CAL package is now permanently installed on the hard disk in the C:\OPSCON directory and can be run from that directory in future by using the DOS command OPSCON, i.e.

C:\OPSCON > OPSCON

Installing OPSCON to start with an icon within WINDOWS 3.1 and WINDOWS FOR WORKGROUPS 3.11

Having installed OPSCON to run under DOS from the C:\OPSCON directory of the hard disk using the previous set of instructions, for those readers operating mainly in a WINDOWS 3.1/3.11 environment, to create an icon within the WINDOWS Program Manager so that OPSCON can be run simply by clicking on the selected icon, use the following instructions.

1. Within WINDOWS maximise the Program Manager and then (using the mouse left button)
2. Click on an appropriate Program Group (say DOS Applications)
3. Click on the File menu option
4. Click on the New menu option
5. Within New Program Object confirm that user is creating a Program Item by clicking on the OK button
6. Within Program Item Properties complete resulting dialog boxes as shown in Fig. A2 i.e.
   Description OPSCON
   Command line OPSCON
   Working directory C:\OPSCON
7. Within Program Item Properties click on the Change Icon button
8. Within Change Icon click on the OK button to accept no icons present
9. Within Change Icon click on Browse
10. Within Browse click on the directory C:\OPSCON
11. Within Browse click on the file OPSCON.ICO
12. Within Browse click on the OK button to select the OPSCON icon
13. Within Change Icon click on the OK button to select the OPSCON icon
14. Within Program Item Properties click on the OK button to confirm that the selected OPSCON icon is installed in the selected Program Group

The results of this set of procedures (the details of which appear in Fig.

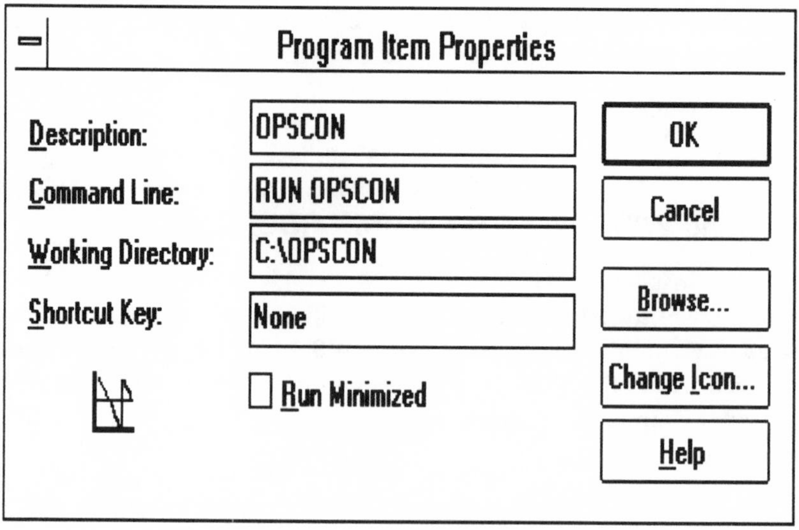

A2 WINDOWS Program Item Properties dialog box completed for creating an icon to run OPSCON within a WINDOWS environment.

A2) should establish the icon within the appropriate WINDOW's program group.

Installing OPSCON to start within WINDOWS 95's Programs Start Menu

Having installed OPSCON to run under DOS from the C:\OPSCON directory of the hard disk using the previous set of instructions, for those readers operating mainly in a WINDOWS 95 environment, add OPSCON to the WINDOWS 95 Programs Start Menu by following the instructions below.

From WINDOWS 95 START button:

1. Select (left click with the mouse) the Settings option
2. Within Settings click on the Taskbar option
3. Within Taskbar Properties select the Start Menu Programs tab
4. Within the Start Menu Programs tab click on Add button
5. Within Create Shortcut type C:\OPSCON\OPSCON.BAT in 'Command line'
6. Within Create Shortcut click on NEXT > button to return to the Select Program Folder
7. Within Select Program Folder click on NEXT > button to move to

Select a Title for the Program and type OPSCON in the 'Select a name for shortcut' box
8. Within Select a Title for the Program click on the FINISH button to return to Start Menu Programs tab within Taskbar Properties
9. In Start Menu Programs tab within Taskbar Properties click on the OK button

This completes the installation of OPSCON as an program option within WINDOWS 95 Programs Start Menu.

To ensure that the DOS Window in which OPSCON runs is closed automatically when exiting from OPSCON under WINDOWS 95, follow the procedure.

From WINDOWS 95 START button:

1. Select the Programs option
2. Within Programs select Windows Explorer option
3. Within Windows Explorer find the Windows folder and expand (i.e. right click on the +)
4. Within the Windows folder find the Start Menu folder and expand (i.e. right click on the +)
5. Within the Start Menu folder find the Programs folder and expand (i.e. right click on the +)
6. Click again on the Programs folder to reveal the names of programs in Start Menu
7. Find the OPSCON program and click with right button (NOTE: must be right mouse button)
8. Select the Properties option
9. Within OPSCON Properties click on the Program tab
10. Within the Program tab of OPSCON Properties click on 'Close on exit' check box
11. Within the Program tab of OPSCON Properties check that name of program in the first box is OPSCON and change if necessary
12. Within the Program tab of OPSCON Properties click on Apply button
13. Within the Program tab of OPSCON Properties click on OK button
14. Close down Windows Explorer

OPSCON's DOS Windows will now close automatically when exiting.

Running the OPSCON CAL package

When running OPSCON as a HARD-DISK USER any files created by the package will be stored in the directory C:\OPSCON. So for most readers who are happy for additional files to be stored in this manner, the required response to the OPSCON's opening screen requesting information as to whether the user is a HARD-DISK USER or NETWORK USER is to hit either the ENTER or H key to indicate that they intend to operate in hard-disk mode.

For those readers wishing to install OPSCON on a network, it is assumed that files created by OPSCON will be saved to a floppy disk located in the A:\ drive. However, to print graphs, save the print or forecast data files created by network users, it is necessary for OPSCON initially to save three small files ERASENOT.PIC, ERASENOT.PRN and ERASENOT.WKB on the user's floppy disk. This initialisation is effected by choosing the NETWORK USER option followed by the NEW USER option. This initialising procedure is only required once, it being assumed thereafter that network users will have placed their initialised OPSCON floppy disk in the A:\ drive of the networked PC.

Having indicated which type of user is intending to run OPSCON, the package's main menu is displayed as shown in Fig. A3. This menu guides the user to either a suite of forecasting packages (FOREMAN) or a suite of inventory control packages and simulation models (STOCKMAN).

OPSCON is designed for ease of use as a menu driven package within which most operations are selected simply by pressing the key representing the first capitalised letter of the required menu option, i.e. Q for Quit but L for coLOUR. For consistency the following commonly used menu words and their subsequent action are described:

EXIT — by typing E for EXIT, a simple Yes/No menu is activated which, if Y is pressed for confirmation, exits the OPSCON package and returns the user to DOS (or WINDOWS)

QUIT — by typing Q for QUIT, a simple Yes/No menu is activated which, if Y is pressed for confirmation, quits from the current OPSCON program and returns the user to either the FOREMAN or STOCKMAN menus.

RETURN — by typing R for RETURN usually infers that the user is returned to the previous menu

CONTINUE — by typing C for CONTINUE usually infers that the user is taken to the next menu

VIEWGRAPH(S) — by typing V for VIEWGRAPH(S) displays either the only graph available or, where several graphs can be selected, displays a menu of the graphs that

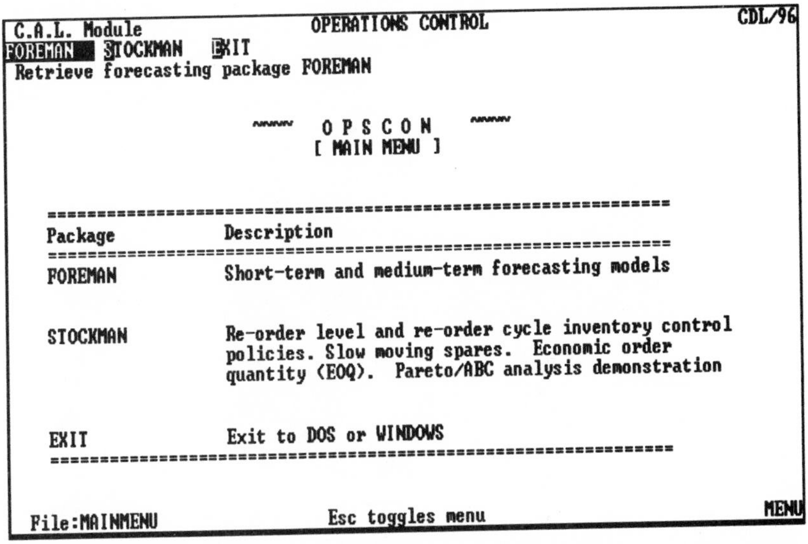

```
C.A.L. Module                OPERATIONS CONTROL                          CDL/96
FOREMAN  STOCKMAN   EXIT
Retrieve forecasting package FOREMAN

                        ~~~~~   O P S C O N   ~~~~~
                               [ MAIN MENU ]

   ============================================================
   Package              Description
   ============================================================
   FOREMAN              Short-term and medium-term forecasting models

   STOCKMAN             Re-order level and re-order cycle inventory control
                        policies. Slow moving spares.  Economic order
                        quantity (EOQ).  Pareto/ABC analysis demonstration

   EXIT                 Exit to DOS or WINDOWS
   ============================================================

 File:MAINMENU                 Esc toggles menu                          MENU
```

A3 OPSCON's main menu providing access to the FOREMAN and STOCKMAN collection of programs.

can be chosen by the user. One option always associated with the VIEWGRAPH(S) menu is:

GRAPHSAVE — which when selected offers the following options:

- NEWNAME — which allows the user to save a .PIC file of the current graph either to the hard disk or to a floppy disk positioned in the A:\ drive (choose Network option)
- OLDNAME — which allows the user to save a new version of an existing PIC file of the current graph either to the hard disk or to a floppy disk positioned in the A:\ drive (choose Network option)
- VIEW — which simply allows the user to view the current graph
- PRINT — which allows users with a suitable printer to print a graph. (See notes to follow on setting up the PICPRT file.)

Further menu options which occur within OPSCON are:

PRINT — typing P for PRINT displays a further menu from which the user can choose:

- PRINT — which then sends selected output to a suitable printer, or

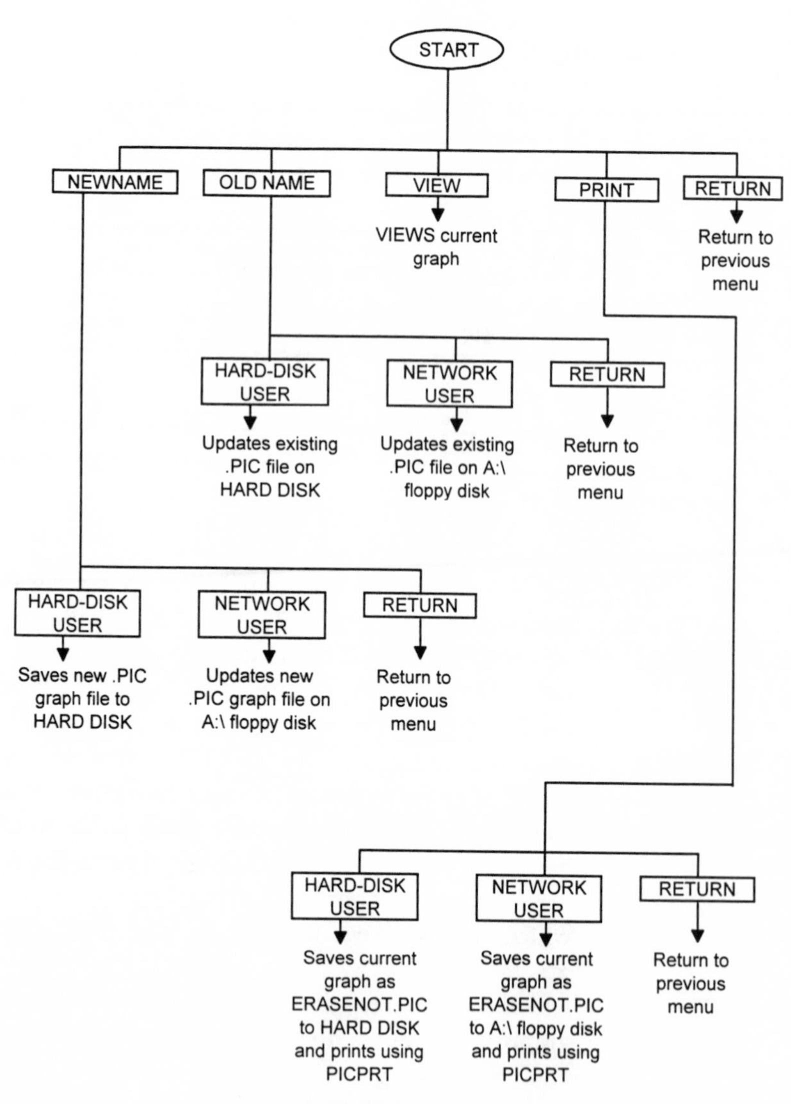

FILE — saves a .PRN file to disk, it being anticipated that the user may subsequently wish to incorporate information into a text document

LCD/coLOUR — which by repeatedly pressing the L character key toggles the screen display between a colour display and a mono (black on white) display, the latter being more suitable for laptop computers with LCD screens

Configuring the PICPRT program for printing graphs

The OPSCON package incorporates a file called PICPRT which allows the user to print many of the graphs generated within the OPSCON package at the time they are displayed as well as by saving .PIC files for subsequent printing. For Lotus for DOS users, the PICPRT file will appear very similar to the PGRAPH facility provided with Lotus 123. To configure PICPRT for a particular printer, from DOS:

1. select the C:\OPSCON directory using CD\OPSCON
2. run PICPRT using the command PICPRT
3. select the CONFIGURE option
4. select the DEVICE option to display a list of suitable printers
5. highlight the required printer by using the space bar and then pressing ENTER
6. select the SAVE option to make the selected printer the permanent, default printing device.

(NOTE: Users with HP Deskjet printers will find that the HP Thinkjet option is suitable.)

Technical support

Technical support for OPSCON can be obtained from the author via E-mail using the address C.D.LEWIS@ASTON.AC.UK.

Appendix B: Simplification of exponentially weighted average formula

Proof to show that a forecast based on a simple exponentially weighted average can be simplified to a statement that:

NEW FORECAST=OLD FORECAST+ALPHA*CURRENT FORECASTING ERROR

If:

$$u_t = \alpha d_t + \alpha(1-\alpha)d_{t-1} + \alpha(1-\alpha)^2 d_{t-2} \ldots$$

then:

$$u_t = \alpha d_t + (1-\alpha)[\alpha d_{t-1} + \alpha(1-\alpha)d_{t-2} \ldots$$

and since:

$$[\alpha d_{t-1} + \alpha(1-\alpha)d_{t-2} \ldots = u_{t-1}$$

it follows that:

$$u_t = \alpha d_t + (1-\alpha)u_{t-1}$$

or:

$$u_t = u_{t-1} + \alpha(d_t - u_{t-1})$$

and since:

$d_t - u_{t-1} = e_t$ i.e. the current forecasting error

it follows that for stationary demand situations, where it is assumed that the forecast for one period ahead is the same for all periods ahead (i.e. there is no growth or seasonality) then:

$$f_{t+T} = f_{t+1} = u_t = u_{t-1} + \alpha e_t$$

or in words:

NEW FORECAST=OLD FORECAST+ALPHA*CURRENT FORECASTING ERROR

Appendix C: Selected values of normal variable, u

Probability of u *not being* exceeded (vendor service level %),
Probability of u *being* exceeded (probability of stockout %), and
Partial expectation E(u).

Normal variable, u	Vendor service level, %	Probability of stockout, %	Partial expectation, E(u)
0.60	72.6	27.4	0.169
0.70	75.8	24.2	0.143
0.80	78.8	21.2	0.120
0.90	81.6	18.4	0.100
1.00	84.1	15.9	0.083
1.05	85.3	14.7	0.076
1.10	86.4	13.6	0.069
1.15	87.5	12.5	0.062
1.20	88.5	11.5	0.056
1.25	89.4	10.6	0.051
1.30	90.0	10.0	0.046
1.35	91.2	8.8	0.041
1.40	91.9	8.1	0.037
1.45	92.7	7.3	0.033
1.50	93.3	6.7	0.029
1.55	94.0	6.0	0.026
1.60	94.5	5.5	0.023
1.65	95.1	4.9	0.021
1.70	95.5	4.5	0.018
1.75	96.0	4.0	0.016
1.80	96.4	3.6	0.014
1.85	96.8	3.2	0.013
1.90	97.1	2.9	0.011
1.95	97.4	2.6	0.010
2.00	97.7	2.3 (2.275)	0.008
2.25	98.8	1.2	0.004
2.50	99.4	0.2	0.002
2.75	99.7	0.3	0.001
3.00	99.9	0.1	0.000

References

1. Brown R G, (1962) *Smoothing, forecasting and the prediction of discrete time series* Prentice Hall, Englewood Cliffs, New Jersey, USA.
2. Harrison M, (1997) *Finite capacity scheduling – the art of synchronized manufacturing*, Woodhead Publishing, Cambridge, UK.
3. Holt C C, (1957) Forecasting seasonals by exponentially weighted moving averages, *Office of Naval Research Memo* No 52.
4. Karush W, (1957) 'A queuing model for an inventory problem' *Operational Research Quarterly* **5**.
5. Kenworthy J, (1997) *Planning and control of manufacturing operations*, Woodhead Publishing, Cambridge, UK.
6. Lewis C D, (1995) 'A purchase quantity decision maker' *Proceedings of the 2nd International Symposium on Logistics* pp 149–154.
7. Mitchell G H, (1962) 'Problems of controlling slow-moving spares', *Operational Research Quarterly* **13**.
8. Shone M L, (1967) 'Viewpoints' *Operational Research Quarterly,* **18**.
9. Trigg D W, (1964) 'Monitoring a forecasting system', *Operational Research Quarterly* **15**.
10. Trigg D W, and Leach A G, (1967) 'Exponential smoothing with an adaptive response rate', *Operational Research Quarterly* **18**.
11. Waters C D J, (1992) *Inventory Control and Management*, Wiley, Chichester, New York and Toronto.
12. Winters P R, (1960) 'Forecasting sales by exponentially weighted moving averages' *Management Science* **6.**
13. Wilson R H, (1934–35) 'A scientific routine for stock control', *Harvard Business Review XIII.*

Index

A items, 73, 134
ABC categorisation, 133
adaptive forecasting, 36
adaptive response rate forecast, 36
administration costs, 13
allocation
  of 'A' items, 73
  'B' items, 74
  'C' items, 75
altering datafiles, 19
annual total cost, 102
average demand, 84

B items, 74, 134
benefits of holding stock, 11
Bill of Material, 4
Brown R G, 45

C items, 75, 135
changing sensitivity, 7
coefficient of determination, 59
cost characteristics of policies, 134
cost sensitivity of EOQ approach, 99
costs
  administration, 13
  operating, 12
  ordering, 13
  stockout, 13
  storage, 13
creating data files, 19
curve fitting, 54
customer demand, 8
customer service level, 111, 124
  'P', 114

CYCLEPOL file menu structure, 128

datafiles, creating and altering, 19
DATAHELP, 19
de-seasonalising factors, 61
decomposition, 61
delayed adaptive response rate forecast, 38
dependent demand, 4
discount, 102
DOS, 144
double smoothed model, 45

Economic Order Quantity (EOQ), 94
engineering spares, 137
EOQ_EVAL file structure, 106
EOQ_SENS file structure, 107
establishing the replenishment order size, 94
EWA file, 25
  menu structure, 42
EWC, 29
exponential growth curve, 56
exponential smoothing constant, 29
exponentially weighted average, simplification, 152

files in OPSCON module, 18
fitted model, 8
fitting, 7
forecast horizon, 7
forecasting, 3, 5
  for growth, 44
  for seasonality, 44

FOREMAN package, 18

GRAPHSAVE option menu structure, 150
growth characteristics, 9
growth element, 50
growth forecasting models, 44

Harrison M, 5
holding costs, 13
Holt Winters' model, 49
Holt C C, 23, 46
hybrid policy, 136

impulse, 8
independent demand, 3
inventory, 68
  control, 3, 79
    policies, 13
  operating costs, 12

JIT, 4

Karush W, 139
Kenworthy J, 5

leadtime, 84
LEVELPOL file menu structure, 92
Lewis C D, 102

management by exception, 33
MAPE, 27
  values of, 28
Master Production Schedule, 4
maximum stock level, 14, 121, 137
Mean Absolute Percentage Error, 27
Mean Squared Error, 27, 55
medium-term forecasting, 7
Mitchell G H, 139
monitoring forecasting systems, 33
monitoring procedure, 6
moving average, 23
MRP, 4
MSE, 27

normal distribution, 153

operating costs, 97
OPSCON CAL module, 5, 18
  installing, 144
  instructions, 148
ordering costs, 13, 94
overshoot, 110

Pareto, 68, 133
  analysis, 69
  approach, 68
PARETO file menu structure, 77
partial expectation, 112
penalty cost, 84
PICPRT file, configuring, 151
POLY2_3 file menu structure, 66
polynomial equations, 58
prediction, 5
procurement leadtime, 81
purchase/acquisition costs, 91, 102

R, 59
re-order cycle policy, 14, 117
re-order level, 14
re-order level policy, 14, 81
regression techniques, 54
replenishment orders, 14
replenishment quantity, 14
review period, 119
ROL policy, 81

(s,S) policy, 136
seasonal characteristics, 9
seasonal element, 50
seasonal forecasting, 48
Shone M L, 39
short-term forecasting, 6
simple exponentially weighted average, 21, 24
SIMPLE file, 29, 41
  menu structure, 43
slow moving items, 138
smoothed error tracking signal, 33

SPARE file menu structure, 142
standard deviation of demand, 69, 84
stationary demand, 8, 21
stationary element, 49
Status of all Materials, 4
step change, 8
stock holding, benefits, 11
stock-on-hand, 14, 81, 119
stockout costs, 13
storage costs, 13, 95
straight line forecasting model, 55
STREXPO file menu structure, 66

time periods, 22
total cost, 102
TREND file menu structure, 52
Trigg D W, 33
Trigg D W and Leach A G, 37
turning points, 58

unequally weighted moving average, 23

vendor service level, 87, 110, 122

Wilson R H, 94
WINDOWS 3.1/3.11, 145
WINDOWS 95, 146

zero demand, 28
Zero Inventory, 4